HOW FITZPATRICK AND MELANIN INFLUENCE THE COSMETIC TATTOO:
A Useful Guide for the PMU Artist

ROSE PRIETO
CPCP, LE, CCE

KDP Publishing

2020 by Rose Prieto

 For permission requests, speaking inquiries or content partnerships, email us at

Rose@BeautyandBrowLounge.com

All rights reserved. No portion of this book may be reproduced, stored in a retrieval system, or transmitted in any form or by any means (electronic, mechanical, photography, recording, scanning, or other) except for brief quotations in critical reviews or articles, without the prior written permission of the author, Rose Prieto.

Self published through KDP

Cover Design: Mark Anderson, Big Whiskey Art (BigWhiskeyArt.com)

Interior Illustrations: designed by Rose Prieto

Interior Illustrations: created by Mark Anderson, Big Whiskey Art

Photograph of Rose Prieto for back cover: Abby Tuano, Life Films

Image of Bellar, Rose Gold and Vertix Needle Cartridge: Courtesy of FK Irons (FKirons.com)

Image of Tina Davies microblade: Courtesy of Tina Davies (TinaDavies.com)

ISBN 978-1-734519204 (paperback)

Printed in the United States of America

This guide is dedicated to the PMU artist who is on a perpetual quest for knowledge. It is a privilege to be part of your journey!

Contents

INTRODUCTION vi
CHAPTER 1: IN THE SKIN 1
   – TOOLS FOR THE AESTHETIC ARCHITECT 3
CHAPTER 2: WHAT IS MELANIN? 13
   – A PIGMENT CALLED MELANIN 15
CHAPTER 3: DISSECTING MELANIN 27
CHAPTER 4: THE SKIN IS OUR BLANKET OF LIFE 37
CHAPTER 5: DO YOU SPEAK FITZPATRICK? 52
   – TERYN DARLING'S DARK SKIN DISCOVERY 57
   – MELANIN IS AN ADDITIONAL PIGMENT 67
CHAPTER 6: WHY DOES POST-INFLAMMATORY HYPERPIGMENTATION MATTER IN TATTOOING? 79
CHAPTER 7: MELANIN INHIBITORS 87
CHAPTER 8: MELANIN'S INFLUENCE ON MICROBLADING 97
CHAPTER 9: POST-WOUND PIGMENTATION AND THE MYSTERIOUS MELANOCYTE CELLS 106
   – THE MYSTERY OF THE FUNKY-HEAL 111
CHAPTER 10: SKINCARE FOR THE HEALING TATTOO 120
   – DEHYDRATED SKIN 124
   – ASSESSING THE SKIN 126
   – DRY SKIN VS. DEHYDRATED SKIN 128
   – LOCKING MOISTURE IN 130
FINAL FOOD FOR THOUGHT 135
GLOSSARY OF TERMS 138
ABOUT THE AUTHOR 153

INTRODUCTION

We delight in the beauty of the butterfly, but rarely admit the changes it has gone through to achieve that beauty. -Maya Angelou

This guide is for the cosmetic tattoo artist that is passionate about truly learning the dichotomy of permanent makeup (PMU). The facets that contribute to a beautiful artistic result require an understanding of art as well as an understanding of skin-related science, yet the latter is a notion that artists are not typically accustomed to wrapping their heads around. Permanent cosmetic artists are taught that micropigmentation is an art, not a science, so how can an artist become enlightened by this concept? My goal, as an artist myself and as an educator, is to change the narrative of the PMU landscape. I welcome the day when higher education trumps fads and trendy techniques because a well-educated artist is an empowered one. This is the first book in my *Useful Guide* series. My goal is to educate and enlighten the artist that is searching for answers. Education does not care where you came from; it only cares about where you are headed. In this book, I will set the tone by speaking in the skin-vernacular, using professional skin-related terms you may not be accustomed to seeing or hearing. I realize that not all artists are licensed skin care or medical professionals; therefore, I have provided you with a glossary of terms located at the end of this guide. In most cases, I will also be explaining the meaning of certain terms as I use them in context; however, I will always emphasize these important words with *italics*. If you'd like to learn more about the mysteries and complexities of the skin after you finish this book, I welcome you to read the second book in my *Useful Guide* series, *In The Skin*.

I hope that the information here will provide you with a better understanding of the arcana of human skin with regard to the hows and whys pertaining to *Fitzpatrick* type and how this scale is related to *melanin*, the quintessential influencer in the skin-healing arena. My goal is to help you to decipher the mystery of the proverbial funky heal. Skin is a living, functioning organ. When the skin is compromised, in any way, the healing cascade is ignited and may influence the integrity of the permanent implant. PMU is a beautification technique, but it is also a minimally invasive procedure that compromises the integrity of the skin (the *epidermis* and *dermis*). When this happens, the hierarchy of healing cells send their special forces to the scene of the injury to heal the wound that was just created by the PMU needle. What happens to the skin cells after the needle hits the skin? How does this affect the pigment implants? What happens to the pigment once it has been implanted and how does this affect the credibility of your work? Will the pigment change color? If so, when, why and how? These are all important questions that I will be addressing within this guide. If you are an artist who has seen or experienced a "funky heal" (a gray-toned or muddy-healed color tone) you have witnessed the mysteries and complexities of melanin first hand. In this book, I will be referring to permanent makeup as *PMU*, *pigment implants*, *micropigmentation* as well as *tattooing* and *cosmetic tattooing*. As PMU artists, we use many names to describe our art, but passion is the one common denominator that we all possess!

A mastery pertaining to the *histology* and the *physiology* of the skin is not required to become a tattooist, but a good understanding is essential and in my opinion just as important as practical PMU experience. A determined artist will spend hours "practicing the hand" (as artists like to say) in an effort to master the needle in live skin. A great deal of emphasis is placed on advanced classes that teach various techniques like ombré shading, hair-strokes, pixel-shading, combo-brows, and powder-brows to name just a few. Artists become seduced by the latest and greatest specialty courses instructed by popular celebrity trainers teaching their fad and technique. Some artists, however, fail to recognize the missing link that is hindering them from excelling. I understand this because I was one of those artists!

Over the years, I have spent thousands of dollars on advanced-training courses which I truly believed I needed in order to become better at my craft. Although each advanced course I attended taught me a great deal of practical experience, the healed work I was generating was uninspiring and left me wanting to become better. In other words, the myriad of advanced training I took helped me to perfect my technique; however, my healed work (specifically eyebrows) was coming back lackluster, and I was often surprised to see new, unwelcomed color tones in my healed work. What was I doing wrong? I had an artistic story to tell that would fail to translate on the skin. I was becoming increasingly frustrated as an artist, not to mention financially drained.

Then one day I realized that I possessed the missing link all along. I had been serving my clients for years as a licensed aesthetician and electrologist. I was well educated within the wheelhouse of skin and skincare, so like Dorothy who discovered that she had the solution to her problem all along (in the form of ruby slippers), I reached within my knowledge bank to search within myself. I realized that the foundation of my business held the answers to my PMU problems. I was educated in matters of the skin yet focused all of my energy and time on practicing the hand. I ignored the information I already knew about the physiology and histology of the skin and failed to incorporate that essential knowledge into my work. Why did I do that? I did that because as PMU artists, we typically compartmentalize art from science, and as I mentioned earlier in this introduction, cosmetic tattooing is dichotomous.

The advanced courses I enrolled in, in my early PMU years, were invaluable to me with regard to developing my practical, artistic technique as a tattooist. I was fortunate to seek the finest master artists to enlighten me about the magic of their technique. As a working electrologist and esthetician, I was well educated about the *integumentary system,* and I often wondered why such little emphasis was placed on the histology and the physiology of the skin. I realize now that it was not the master trainer's job to teach me about the eccentricities of the skin. These elite educators were masters at moving the needle, but teaching the artist about the body's largest organ was not in their wheelhouse. As an esthetician, a thorough understanding of the epidermis is critical due to the many modalities that can cause trauma to and/or pigmentation concerns on the skin. The study of aesthetics requires the practitioner to have a detailed understanding of the complexities of the skin they are treating, even though the treatments are always superficial, unlike tattooing which occurs in the live tissue, the *dermis.* The question is, why doesn't the cosmetic tattooing arena hold the same standards with regard to higher education pertaining to the skin? Advanced training courses typically do not emphasize the importance of the integumentary system, and more importantly, how that system works against the pigment implant. This is a huge conundrum for me. As a young artist, I compartmentalized the two topics and assumed that PMU belonged in one box and the mysteries of the skin went into another. Master trainers continued to teach the latest technique and I was thrilled to be learning from them. Why was it, however, that my work was still coming back dark and gray when my intention was light and warm? I did not use dark, gray tones when working on my client, so why was this happening? The master-trainers I learned from knew how to teach the hand brilliantly, but what about the proactive measures I could have taken to ensure a predictable heal?

The PMU industry is driven by beautiful artistic results. Instagram-ready work is appealing to the audience, yet these photos are often photoshopped or taken immediately after the procedure, and they often fail to showcase the healed work months or even years post-procedure. The notion of continuous practice on skin in an effort to create beautifully executed, visually pleasing results is what artists identify with. I agree with this philosophy; after all, they say that "practice makes perfect." I question, however, whether it takes 10,000 hours of practice (as Malcolm Gladwell explains in *Outliers)* for a PMU artist to truly become great. I've witnessed many spectacularly talented new artists generate amazing work, so I would disagree on that specific number. There are artists who have been moving their needle for 3,000 hours or even 1,000 hours whom I would consider to be very talented. Artistic talent is innate and new artists inspire me every day! It is my belief, however, that this "10,000-hour" theory has more to do with the time it takes to learn and digest every intellectual nugget that an artist must know, with regard to education pertaining to the skin. Yes, practical technique improves over time and by the time an artist achieves their 10,000-hour mark, they have taken their skin to a more elevated level. But what about that conundrum I alluded to a moment ago? Those questions can only be answered over time through advanced education and by digesting critical information like the nuggets I am offering here.

I have been a makeup artist and an eyebrow-design artist for nearly three decades and run a thriving beauty business in Miami named *Beauty and Brow Lounge*. Having mastered the art of facial contouring, as a theatre arts major in the early 1990's, my artistic focus was playwriting and stage makeup. At an early age, I learned that aesthetic beauty is all about math; it's about balance and proportion. I learned how to manipulate the contours of the face, using shadows and light, based on character analysis and character development. I learned to create contours that changed expressions and appearance. The ability to manipulate the features of the face can be game-changing for the individual who longs for aesthetic change.

The art of contouring is indeed powerful! For example, if a woman's eyes appear droopy at the outer corners (lateral canthus) giving her a "sad puppy" appearance, the PMU artist can strategically implant eyeliner in a way that will give her eye the illusion of lift. As a makeup and contouring expert, I had always relied on makeup to create an illusion. However, when I became a cosmetic tattooist, I realized that it wasn't simply math that contributed to aesthetic balance, but science as well. The ability to change lives for the better by creating a better version of a person's facial features is an honor. As cosmetic tattoo artists, we strive to make a positive change by manipulating the shadows and light of the skin.

Tattooing is, however, a finite process. As artists, we create an opening within the vascular portion of the skin (the *epidermal/dermal junction*). We implant pigment into that vascular depth, and then we wait for the skin to heal. For example, when the tip of a finger becomes exposed to a paper-cut, the recipient feels pain and can even experience bleeding. The superficial cut from a piece of paper is not very deep and does not require medical attention. However, the depth of a paper-cut is similar to the vascular depth for which a PMU artist implants pigment; in fact, at times, a paper-cut is even deeper than a PMU pigment implant! On paper, tattooing is a simple process. Put simply, it is the act of implanting material within living tissue and then allowing the skin to heal until the implanted color is revealed to the naked eye. However, for those new and seasoned artists alike, the process is anything but simple. The art of micropigmentation is complex and requires know-how in various areas of knowledge, such as color theory, skin anatomy and histology, skin type and skin assessment, not to mention needle knowledge, practical experience working on skin, as well as a myriad of other areas necessary for the artist to master the craft.

This guide is dedicated to the artist that has yet to find their groove and is striving for greatness within the PMU arena. The knowledge and practical skills that are needed to become a well-rounded PMU artist takes time to digest. I hope that the information I have provided here with great passion and love for the PMU industry, helps you to get closer to the top, as you propel yourself toward being the best artist you can be.

CHAPTER 1: IN THE SKIN

Human beings are actually encased in a seamless raincoat of dead cells known as the stratum corneum. Your stratum corneum represents the final stage of the life cycle of the cells that make up your epidermis or outer layer of the skin. The life cycle of epidermal cells begins at the interface between the epidermis and the dermis.

Arthur K. Balin, M.D., Ph.D., and Loretta Pratt Balin, M.D., "The Life of the Skin: What it hides, What it Reveals, and How it Communicates"

As a cosmetic tattooist, esthetician and electrologist in Miami, Florida, I am truly blessed to serve skin types of all races, textures, colors, and ethnicities. That honor provides me with a unique perspective into the eccentricities of the many different skin types. It is not lost on me that skin is not one size fits all but rather a living, breathing, well-oiled machine, and no two are alike. The skin is an organ of entry. It is a super-machine known as the *epidermis*, the non-vascular region of the skin made up of complex cells. Each skin-related cell is programmed with a specific task in an effort to maintain the skin-machine's efficiently. The human body has the ability to physically achieve amazing things. It can construct magnificent buildings, it can endure a mountain-climb, it can compose sweet music, and it can create awe inspiring masterpieces of art like Michelangelo's *Pieta* which always brings me to tears each time I view it up close. The human body is truly epic, to say the least. However, with all its ability and intelligence to create whatever it chooses to create, the human body would cease to exist without the armored protection of its largest and most elegant organ, the skin, part of the *integumentary system* of the body.

When we think of the permanent makeup beautification process, many artists fail to recognize how complex that undertaking truly is. There are so many more factors involved than simply introducing the needle to skin. For artists, permanent makeup is a creative process that is viewed from their particular artistic point of view. As artists, it is our passion to create something more beautiful than it previously was. However, what many of us fail to comprehend is the influence that science has on that artistic process. In my opinion, performing micropigmentation on a client without a clear understanding of the complexities of the skin is like navigating a ship through open waters without a compass. In the artist's defense, a license in aesthetics or medicine is not a requirement to practice cosmetic tattooing, so that is why skin-care nerds like me are keen on enlightening and educating the PMU community, in an effort to change the narrative and empower artists through proper education.

- TOOLS FOR THE AESTHETIC ARCHITECT

When I first meet with a client, I immediately assess their skin. I examine all the markings on the face, such as freckles or *post-inflammatory hyperpigmentation*. I look for any signs of *hypopigmentation*, *melasma* or any other obvious skin condition. I look at whether they have under-eye circles, and if so, what color-tone is this under-eye discoloration? Is the under-eye pigmentation brownish-black or reddish-brown? Are the circles barely there, or are they deeply saturated, giving the client a black-eye effect? I look at the client's eye color as well as their hair color and texture. I look at their brow and lash growth patterns; is the hair thin and sparse, or is it thick and dense? Thick, dark, dense hair is an important identifying clue with regard to ethnic influence.

I take notes about any scars I see, and if there are no obvious scars, I ask if they have any. Then I ask probing questions about each scar: How did the the scar get there? How long has the scar been there? Did they treat the injured skin in an effort to prevent a scar from forming, and if so what was the protocol they used during the healing phase? The final question I ask is, "Are you typically prone to scarring, regardless of the placement on the body or the type of injury?" I examine the color and texture of any existing scar and I note any possible *keloids*. I observe if the scar is light pink, dark brown or raised. I pay attention to old scars, how they evolved and if they faded almost entirely. A scar can tell you a great deal about the genetic healing-code of the client. If a client is peppered with hyperpigmented marks on their arms and face, that is a red flag for me. It tells me that their skin may be vulnerable during the PMU procedure, resulting in a compromised color-heal (see chapter 5). Let's take the skin surrounding the lips, for example. The lips are indeed unique and differ from the skin on the face, as they are void of *keratinocyte cells* as well as *hair follicles*, but what about the lip's surrounding skin; the keratinized skin that lies just beyond the fringe? When meeting with a new client, examine their lips. Note the small patch of skin on the outer corners of the mouth, as well as the skin just above their vermillion border (the "milk moustache" zone). Are these areas the same tone as the facial skin or are these areas dark and richly pigmented? If the skin within these surrounding areas is darker than the rest of the face and body, it is a red flag that this client is prone to hyperpigmentation and even hypo-pigmentation. When I see overly pigmented areas surrounding the lips, particularly within the areas I've just mentioned, I immediately ask the client probing questions about their lifestyle: Do they wear lip plumper often? Do they drink acidic beverages often (such as, juices with lemon, orange or even pineapple)? When exposed to UV radiation, skin that is peppered with irritating agents, such as acidic residue from a beverage, will hyper-pigment very easily. Probing questions like this are essential and the artist must learn to identify outlying areas for pigmentation clues, before attempting to neutralize the cool undertones of melanin-rich lips. Note that individuals who have a tendency to hyper-pigment on or around the lips as a result of external influences, may also experience pigmentation issues in the axillary region (under the arms) or even between the thighs, as a result of friction. An artist with a keen eye who looks outside of the box for skin clues, will have a much better understanding of the skin they're tattooing in, then the artist who wears blinders and chooses not to look behind the face for evidence.

After viewing the surrounding skin of the lips, examine the lips themselves and note the color. Are the lips pink, light red, light peach, flesh-tone or are they brown, brown-black or even purple? Examine the saturation of the lips as well as the texture. If the lip pigmentation is brown, how saturated is that brown? Are the lips dark as a result of post-inflammatory hyperpigmentation or genetics, or do they have a combination dark lip as a result of both hyper-pigmentation and genetics? Can you even identify the difference between the two? Book 2 of my *Useful Guide* series will go into more detail about the unique skin of the lips and how to identify the two.

After I examine the lips, I look to see if capillaries are visible anywhere on the skin. I look for *couperose* skin, which is skin influenced by dilated blood vessels, particularly on the cheek and nose region. Couperose skin is typically weak in elasticity and would be labeled as "sensitive skin". I always take note if the client has excessive blood vessels on the eyelids, and if there is any veining on the skin. I observe the color of the veins: Are they red, blue or green? When I touch the client's skin firmly, what kind of response do I get? Does my finger leave a red imprint on the skin indicating sensitivity and if so, how long does the imprint linger? I ask the client detailed questions about how their skin responds to sun exposure. I sometimes even ask to see photos of the client as toddlers, this tells me a great deal about the evolution of their skin tone. In addition, viewing photos of your client as a toddler will give you the genesis of their personal color-story. Your client may currently have dark hair and fair skin, but it is important to note if their toddler photo revealed platinum blond hair and golden skin.

Look at the client's facial features and try to form an educated guess regarding their ethnic origin before asking probing questions about their ancestry. This is personal information indeed; however, identifying your client's ethnic origin will assist you in determining how their skin will respond to the tattooing procedure. (I will go into detail about this later on in this guide.). As an aesthetic architect, I need to know as much as I can about the ingredients of the client's "ethnic soup", so to speak. I must get a grasp of the skin sitting in front of me. Upon analyzing the client's skin, and after asking detailed questions about their ethnic origin, I ask myself an important question that will make or break the execution of my work: "With what I now know about this client's skin and ethnic background, how will melanin affect their healed tattoo? What proactive measures can I take to ensure that the pigment implants' healing is as true as I can expect?"

Practitioners are not typically trained to visually dissect the client's skin type. For the most part, the majority of emphasis, with regard to client analysis, is placed on identifying the client's facial structure, hair color, and basic skin color. Within the PMU arena, the artistic vision as it pertains to an eyebrow, lip, or eyeliner tattoo is typically the primary focus. The histology and the physiology of the skin is typically not the main concern for the artist. In most cases, an organic, creative, artistic process is not cohesive with science, because, after all, cosmetic tattooing is an art, not a science, right? Or is it a science too? In my opinion, it takes a mastery of both to become an elite artist.

Earlier I mentioned the word *ethnicity*, which is an enormously important component in the PMU process, if the cosmetic tattooist wishes to see a predictable and true heal, whereby the integrity of the chosen color has manifested in the healed work. In other words, do you think artists and clients desire a funky-gray or purple heal, or a final heal where the selected color is revealed as a rich, warm end result? The color swatched on the client's skin does not always translate into the final healed-result, and (in my opinion) translating the integrity of the chosen color, through the healed tattoo, is the holy grail of permanent makeup! In fact, the word "predictable" should not be used in the PMU vernacular. Predictable is unrealistic when working on living skin. Skin is complex and is not one size fits all. Having a dialog with the client about their expectations is essential and will set the tone for realistic expectations.

PMU can be a crap-shoot of sorts unless the licensed professional has a thorough understanding of how the skin works and why the skin responds to injury the way it does. Why do I mention the word *injury*, and why is the action of "causing an injury" relevant within this topic? It is relevant because permanent makeup involves causing an injury to the skin, a controlled injury, but an injury nonetheless. There are many beauty-related procedures that injure the skin in a controlled fashion for the purpose of beautification. Some examples are fractional CO_2 laser-resurfacing (*Fraxel*), micro-needling with RF, micro channeling, chemical peels, IPL, and PRP injections, just to name a few. Within the world of skincare, injuring the skin in a controlled manner for the purpose of creating "new skin", is common practice. If the practitioner does not have a solid understanding of the ramifications that may occur during the aforementioned procedures, including PMU, he or she is not serving the best interest of the client. At the end of the day, that's the goal, isn't it? Our goal, as professionals, should be to serve the best interest of the client in front of us.

CHAPTER 2: WHAT IS MELANIN?

What, is the jay more precious than the lark because his feathers are more beautiful? Or is the adder better than the eel because his painted skin contents the eye?

-William Shakespeare, The Taming of the Shrew

Have you ever roasted marshmallows over a fire? What happens to that white fluffy confection once the flame hits it? It starts to turn yellow, then deep gold, then brown, until it eventually chars to a black crisp. The skin responds similarly. The marshmallow is to the skin what the colored-response is to melanin. When the marshmallow is faced with an injury (the fire), it's color changes.

Melanin is the tint that gives skin, eyes, and hair their unique color. In addition, melanin is also part of the skin's defense mechanism and appears anytime the skin is faced with trauma of any kind. When pigment is implanted within the dermis during the tattooing process, melanin will eventually creep its way into the healing wound. Although PMU is a beautification process, it creates a wound within the skin and eventually that wound must heal. Whenever the skin becomes compromised, melanin responds by setting up camp at the location of the damage. This "tinted response to injury" is about to contaminate the beautifully executed, pixel-shaded eyebrow that was just implanted! In other words, the melanin pigment that is produced by the body that has either a yellow, brown, or black cast to it (depending on the individual) will inevitably become added to the cosmetic tattoo that was just implanted within the skin!

Melanin will, without fail and without an invitation, influence the integrity of the tattoo. It will initially visit the site of injury, will linger around for a while as the injury heals, and can then in some cases choose to never leave. When melanin chooses to linger within the healed wound, the skin's tone changes and, by extension, changes the color of the healed pigment implant.

So what is a PMU artist to do? You must first understand that the singular role of melanin is to serve and protect the skin. It does not care about your artistic creation, it only cares about protecting the largest organ of the body. How can we, as aesthetic architects, maintain the sincerity of the art we intended to create? To answer that question, I must first explain the why, when, and how of melanin's influence on the tattooing process. A thorough understanding of why and how melanin influences an injury is the ultimate goal.

- A PIGMENT CALLED MELANIN

The initial creation of melanin occurs during a process known as *melanogenesis* which literally means "the creation of melanin." Melanin is a pigmented biopolymer produced when melanogenesis takes place directly within the *melanocyte cells* (see Figure 6). Melanin is a major contributor within the PMU healing phase, and the human body cannot experience an injury of any kind without its influence in the healing process. Melanin is to skin what armor is to a knight, but it also serves other functions, three in total. Its primary function is to protect the skin by absorbing UV light, it's second job is to protect the skin if it becomes compromised by an injury, and its final function is to provide the skin, eyes, and hair with their unique hue. The word *melas* is Greek for black and is the root word of many pigment-related terms, such as *melanoma*, *melasma*, and *melanosome*, and all of these examples have definitions pertaining to something black (or dark) associated with the skin. In fact, the phenomenon of melanin appearing on the skin during an injury is an immune response!

Melanin is elegantly produced within a sort of "pigment bubble factory" called *melanosomes*. Melanosomes are like pigment transport vehicles that are produced by *dendritic cells* known as *melanocytes*. It is important to understand that dendritic cells are cells that directly support the immune system, which emphasizes the immune-response role these cells play within the skin. I want to take a moment to talk about the dendrites of a melanocyte cell because I find them fascinating! Melanocytes are stationary cells that live within the *basal layer* of the skin (the deepest layer of the epidermis, just above the *epidermal/dermal junction*, see figure 1). I will go into more detail about the layers of the skin in chapter 4. Every human, regardless of race, has (more or less) the same number of melanocyte cells, which (to me) look like an upside-down octopus, with its bulbous body facing south and it's dendritic arms reaching north. The melanocytes produce melanin within their cell-factory, but because these cells are stationary, they need an elegant process to deliver this tinted enzyme to the surface of the skin. The melanocytes' dendrites are like transport highways that melanosomes use to carry and deliver melanin to the *keratinocyte cells* (I will talk more about this in chapter 6).

Once the delicate melanosomes bubble travels out of the dendrites, they are not stable enough to travel throughout the skin without assistance. They require a transport system to help with the distribution of their melanin-pigment cargo, so the melanosomes deliver the melanin supply to neighboring *keratinocyte cells* (the final destination) as a result of either UVR (ultraviolet radiation) exposure to the skin or during the wound-healing phase of an injury. What melanin does, once it has hopped onto the keratinocyte cell, is to blanket its nucleus, forming a protective melanin-umbrella over that nucleus.

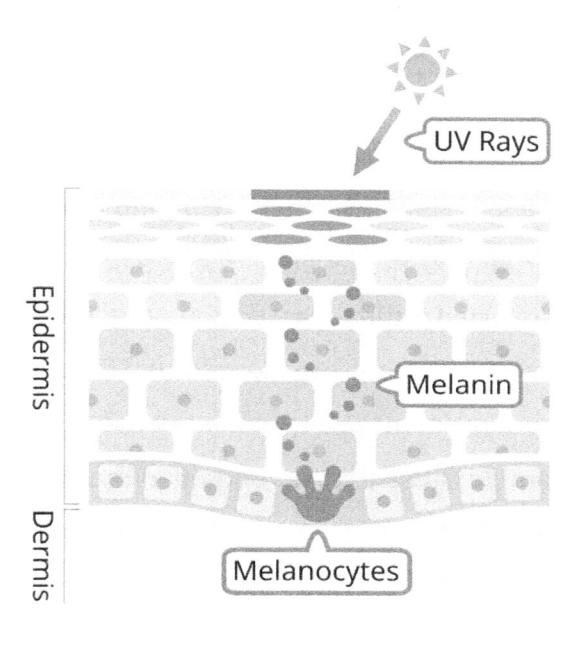

Figure 1

Have you ever seen footage of an airplane spraying gallons of water over a massive forest fire? That is the image I always think of when I picture melanin's response to a surface injury like sun exposure. Like the water sprayed from a plane, it provides a protective covering onto the surface of the skin in an effort to shield it from further damage.

The skin is a self-healing super-machine that has elegant systems in place to proactively execute a defense plan in an effort to avoid getting injured, and that's the takeaway here. The skin is an entity that actually keeps us alive. It protects the body from external influences that would otherwise kill us. The moment the skin comes face to face with a potential threat (UV radiation from the sun, for example), it sends melanin filled melanosome bubbles to the keratinocyte cells. I like to think of skin defense as a team of soldiers that acts in synergy to serve and protect our skin. Melanosomes are not stable and can only travel so far. They rely on the keratinocyte cells to complete their transport mission.

After melanin has been delivered to the keratinocytes, the tint is then deposited to the epidermis, melanin's resting place. The melanin transport cells act like firefighters hosing down a burning building, similar to the plane scenario. In this example, the transport cells spray the skin cells with melanin liquid armor in an effort to shield the skin from UV damage that would otherwise harm the epidermis in the form of a burn (think of that charred marshmallow). The act of coating the skin with melanin creates a protective barrier on the surface, which we physically identify as a suntan. Whenever you see skin that is kissed with a tan, you are witnessing the skin's "serve and protect" response first hand.

I've been mentioning the *keratinocyte cells* quite a bit, so it is only fitting that I tell you more about the importance of these unique cells as they relate to melanin. Keratinocytes live in the outermost layer of the skin, known as the epidermis which is primarily made up of these cells; in fact the epidermis is composed of (upwards of) 90% keratinocyte cells making them the primary resident of the epidermis. These cells are born in the basal layer and eventually work their way up to the outermost layer, creating a super-efficient barrier that protects the body from pathogens, bacteria, viruses, as well as UV damage and water loss. Keratin is the main protein present in these super-cells, which assist in the formation of the cell's cytoskeleton. Hair and nails are also composed of the keratin protein, but the skin's acid mantle lipid barrier keep the cells, and by extension the skin, from cornifying thanks to the precious oils produced from the sebaceous glands. For the record, the skin on the lips do not contain keratinocyte cells or sebum-producing hair follicles for that matter, which accounts for the thin, soft, cushy skin of the lips. This also explains why the lips dry-out, as a result of the void in sebum production.

The keratinocytes are arranged within four separate sedimentary layers within the epidermis: the stratum basale (the basal layer), stratum spinosum (the prickly layer), stratum granulosum (the granular layer), and stratum corneum (the horny layer). As I mentioned earlier, every human possesses the same number of melanocyte cells, what differs is the structure and function of their particular melanocyte cells as well as the amount and type of melanin produced within each cell. This phenomenon is 100% based on genetic influence (read more about this in chapter 3). Melanocytes within Fitzpatrick skin types 5 and 6 are larger and more dendritic than in the other Fitz types. Aside from that, we all possess the same number of these stationary supercells.

Aside from protecting the skin from UV damage, melanin is a pigment that is found primarily in the skin, hair, and eyes. In a nutshell, it is the dye that provides pigmentation to these areas, and the pigment that influences the color of living skin. To be specific, melanin is the dye that supplies the *epidermis* (non-vascular outer layer of skin) with its unique color density; the *dermis* (live/vascular layer of the skin) is not influenced by melanin and is the same color in all humans, regardless of race. The dermis is the vascular portion of skin where artists implant pigment during the tattooing process. What distinguishes humans from one another, is the influence that melanin has to our epidermis and its intensity within the skin; the epidermis is the outer skin that protects our living tissue from the outside world.

When I mention "intensity", I am referring to the unique distribution of color specific to each individual; think of the glass half-full, half-empty theory. Some people have their glasses filled to the brim with melanin, and some have barely a quarter of a glass filled. A person born with albinism, for example, will have some type of pigment-thumbprint to their skin, although they are born without the ability to produce melanin. Albinos have a pinkish hue to their skin, which is tinted by hemoglobin; no human is as white as copy paper. This is an important notion to digest. Without understanding the anatomy of the epidermis, why would an artist attempt to tattoo matter within living, ever-evolving tissue? How will the artist know what to expect from the healing process? The primary influence after the skin is compromised (in any way) as a result of an injury, is melanin. The tattooist must be able to understand why melanin is produced and why it exists. This information is crucial for the successful longevity of a pigment implant.

The epidermis serves two purposes: First, it protects the human body from exposure to the outer world. Without a barrier of protection, our live tissue would be exposed to pathogens, contaminants, and disease, resulting in a fatal outcome for humans. Second, the epidermis serves as the "first response team" of the body's immune system, and is influenced by *Langerhans cells,* which are peppered throughout the epidermis; their highest concentration can be found within the *Stratum Spinosum* (see chapter 4). Langerhans are smart. Like detectives, they identify dangerous organisms that can harm the body. Since they are immune cells, they are dendritic and reside within the basal layer of the skin. The role of the Langerhans is to capture foreign intruders and deposit them to other immune cells within the blood; the body then disposes of them accordingly. For the record, the keratinocytes are the cells that initially send a signal to the Langerhans whenever they're needed. Without the keen-eye of the keratinocytes, the Langerhans would never receive their call to action.

The skin as an entity is part of the body's immune system. It is a self-healing, wondrous super-machine whose singular duty is to serve and protect the body!

CHAPTER 3: DISSECTING MELANIN

Nature's first green is gold, her hardest hue to hold. Her early leaf's a flower, but only so an hour. Then leaf subsides to leaf, so Eden sank to grief, so dawn goes down to day. Nothing gold can stay.

-Robert Frost

There are two types of melanin: *eumelanin* (black-brown based) and *pheomelanin* (red-yellow based). Most humans produce more eumelanin, although both types of melanin exist in human skin and hair. The amount and ratio of each type of melanin is based 100% on genetic predisposition. A *Fitzpatrick* 1 skin type with fair skin, red hair, blue eyes and freckles may not contain any eumelanin at all, as opposed to a Fitzpatrick 2 blond with golden skin and green eyes, who will possess a unique percentage of the two types of melanin. Pheomelanin is the red-yellow tint that provides redheads with their unique hue (see chapter 5 for a detailed description of the six Fitzpatrick skin types).

I classify melanin-influence into three components, the first of which already exists in the skin, and determines our natural skin tone regardless of the UV influence of the sun. When a baby is born, its genetic blueprint is already determined and the skin's natural color evolves as per their DNA. The second component is the influence that the sun has on the skin. To determine the Fitzpatrick type of a client, you can ask questions like "If you stand in the midday sun, how much time will pass before your skin starts to burn?" For some, the answer is ten minutes, and for others, perhaps it is two hours. The way an individual's skin responds to ultraviolet (UV) radiation is essential information. An artist must keenly identify the skin they will be tattooing in order to achieve successful results! The third component is how the skin responds to injury and how the evolution of that injury will appear over time. What will that healed injury look like in eight weeks? What will it look like in two years? When we talk about injury, some may think of a skinned knee, and yes that is indeed an injury. However, so is a bug bite, a sunburn, a pimple, a burn from a chemical peel, a tattoo and a suntan. Although visually appealing, a suntan is a response to an injury on the skin. Anytime the barrier of the skin is compromised, in any way, the healing cascade ignites, triggering melanin to rush to the scene of the damage, like Navy Seals on a mission.

I like to think of melanin as brown liquid armor, because it actually has a function that expands further than the proverbial Coppertone tan. Aside from giving skin its unique color, melanin is part of the skin's defense team. Its purpose is to absorb ultraviolet light in an elegant effort to protect the skin from sun damage. Earlier, I named the two types of melanin. Although all humans produce the same quantity of melanocytes, these melanin-producing cells do not produce the same cocktail of eumelanin and pheomelanin. All humans may contain both types of melanin within their melanocyte reserve, some may contain a small amount of one and the majority of the other type, and some humans may only contain one type of melanin within their reserve. This occurrence is based on the individual's genetic description. This is why protocols that involve the skin cannot fit into a one size fits all box. Skin is NOT one size fits all! Tattooing the skin of a Fitzpatrick-1 fair-skinned person with red hair, blue eyes, and freckles will render a completely different protocol than tattooing a Fitzpatrick 4.5 Latina with dark hair and dark eyes. Albinos are astonishingly void of melanin; earlier I explained why their skin casts a pink hue, as a result of hemoglobin.

Melanocytes are considered to be anti-inflammatory cells that also play a role as free radical scavengers; they seek to reduce oxidative damage within the skin. Earlier I mentioned that the epidermis serves as the first response team to the body's immune system, so this makes sense.

Once a wound has been ignited and the inflammation cascade begins, the melanocytes ignite the distribution of melanin to the site of the injury in an effort to reduce oxidative damage. The melanocytes have a primary role, which is to ignite the distribution of melanin to the wound succeeding the skin's re-epithelialization phase of wound healing. The question, however, is: How much melanin is the melanocyte releasing into the wound, and how often? This circumstance is an enigma and is unique to the individual. How can you identify skin with hyperactive melanocytes? You can do so by asking probing questions based on the clues that can be identified on the skin already. I personally have hyperactive melanocytes and have the ability to hyper pigment even after the most superficial wound, resulting in brown marks that have remained on my skin for years. That scenario is why every PMU artist must be conservative with their initial approach to tattooing melanin-rich skin. You never know what individual will have hyperactive melanocytes.

Since we are discussing the effects of melanin in the skin, this dye represents varying shades of yellow, red, brown and/or black. An example of melanin in action is the visual of sun-loving beachgoers sipping piña coladas on holiday in Rio de Janeiro. Imagine bikini-clad bodies on a beach and you will picture golden-brown and bronzed skin on many of the beach bods around you. The sun-kissed skin you are imagining has been influenced by melanin production, which literally means that you are witnessing their skin becoming injured right before your eyes! This injury is being caused by the UV radiation of the sun. The pigmentation we call a "suntan" is actually a response to that injury. When the skin is exposed to the UV rays of the sun, melanocyte cells trigger the melanosomes to deliver melanin to the neighboring keratinocytes, who then deliver their melanin-soaked selves to the surface of the skin, in an effort to protect the skin from burn and other UV damage. The sun's UVB rays are the burning rays that can lead to a myriad of skin damage, including skin cancer (UVA and UVC rays are also damaging, but I want to use the example of a sunburn caused by UVB rays, to explain this example).

Darker-skinned individuals, known as *melanin rich*, have a greater eumelanin supply within their melanocyte reservoir compared to lighter-skinned individuals. As I mentioned earlier, all humans possess (more or less) the same quantity of melanocyte cells, what differs is the amount and the type of melanin produced by the individual. The varying degree of melanin levels in the skin is directly related to genetic predisposition. In other words, someone native to Norway will have less eumelanin in their eyes, hair, and skin that someone of Middle Eastern descent would, where historically the Middle Easterner has to endure a harsher, more sun-soaked climate which requires a higher concentration of the darker eumelanin. The individual that is native to Nigeria, for example, will have an even greater percentage of eumelanin in their reservoir. This is a valid notion to consider when selecting the appropriate color for your client because in the U.S. we live in a melting pot of inter-mingled races. It is rare to meet a person of pure Scandinavian descent or someone who is 100% Irish. I myself am a combination of Caribbean-Cuban, Bedouin and Lebanese descendants which would classify me as a melanin-rich woman who tans easily, rarely burns and has the tendency to hyper and hypo pigment when exposed to the sun. The keyword in the previous sentence is the *sun.* The primary focus, when determining a client's *Fitzpatrick* type is their skin's response to UV radiation; this information is key.

A common mistake made by artists is the assumption that fair-skinned individuals possess fewer melanocyte cells than those with melanin-rich skin; this is not accurate. Every human contains the same number of melanocyte cells, more or less. As I've mentioned a few times already, what differs is the quantity of melanin and the type of melanin produced by and within these cells. Is the melanin-glass half full, half empty or overflowing? This is why varying ethnicities are diverse when it comes to skin color and it is that diversity that makes the world so beautiful. We are a melting pot of flavors, a rainbow of endless visual delight, which is probably why I'm personally obsessed with eyeshadow palettes, so many colors, so many color choices, so many color combination possibilities! What if we lived in a world where everyone had pink skin? How boring that would be! The amount and type of melanin produced by the melanocytes is pre-determined as a direct result of genetic predisposition. The ingredients that make up an individual's "genetic soup" are the most important elements in the PMU process. Without that general knowledge, the artist is playing roulette with color in the skin, and navigating without a compass.

How Fitzpatrick and Melanin Influence the Cosmetic Tattoo

CHAPTER 4: THE SKIN IS OUR BLANKET OF LIFE

Two layers compose the skin, the superficial epidermis and, deeper, the dermis. Between is a plane of pure energy where the life-force is in full gallop.

-Richard Selzer, M.D., Mortal Lessons: Notes on the

Art of Surgery

I love the quote printed above. "The plane of pure energy where the life-force is in full gallop" is the epidermal-dermal junction where we, as tattoo artists, implant our works of art. This is the proverbial sweet spot, that precise location where many artists can actually feel when they work in skin. Tattooing becomes poetic when the artist can confidently tattoo within that plane of pure energy; it is the holy grail that new artists yearn to find. I consider the epidermis to be the body's blanket of life, and the dermis is the bed it covers and protects.

Look at the skin on your arm right now. The portion of uncompromised skin that is seen by the eyes is the *epidermis*. Our epidermis is an essential organ that is critical for life. Ranging from half a millimeter to just a few millimeters in thickness, the skin of the epidermis (along with the hair, nails, and exocrine glands) is the largest organ in the *integumentary system* and, by extension, the largest organ of the human body. The skin regulates the body's temperature and provides it with a waterproof super-barrier of defense against the outer world; it actually protects the human body from germs, pathogens, viruses, and diseases that would otherwise kill us. The skin is our body's living shield, our blanket of life.

As a super-machine, it has a built-in skincare factory, called the *sebaceous glands,* located within the hair follicle, which produces an external moisture barrier known as *sebum* (the oil that we see on our t-zone and that greases up our unwashed hair). *Lamellar granules* (fatty lipids) in the *stratum corneum* layer of the epidermis is what gives skin its waterproof barrier. But let us back up a bit: Look at the skin on your arm again. What you are actually looking at is the top layer of the epidermis, the stratum corneum, which is an accumulation of dead cells. The misconception that many have is that the epidermis is non-living; it is not live, but rather, dead cells that shed approximately every 28 days. This statement is false; the epidermis is very much alive. It simply does not contain a blood supply. Also known as the "horny layer" (It is made up of the same keratin as a horn on a rhinoceros's head), the stratum corneum is the outermost layer of the five layers of the epidermis which is in a perpetual state of cell turnover. This is the layer, of the five, that I would classify as non-living due to the fact that its job is to shed away spent skin cells. Note the five layers of the epidermis in Figure 2.

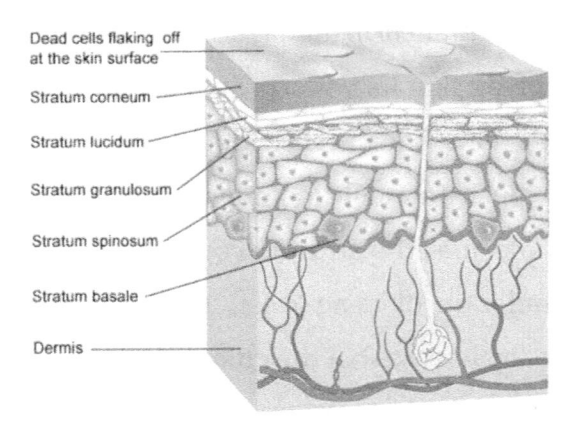

[Figure: Diagram of skin layers with labels: Dead cells flaking off at the skin surface, Stratum corneum, Stratum lucidum, Stratum granulosum, Stratum spinosum, Stratum basale, Dermis]

Figure 2

The top layer, the aforementioned stratum corneum, is a culmination of non-living keratinocyte cells that are waiting to be shed from the body. In healthy skin, the dead cells of the stratum corneum will shed approximately every 28 days, which is why a suntan will eventually fade away. I often use this suntan example with clients, when explaining the process of microblading. Newer, less experienced microblading artists are often misinformed and tell their clients that microblading is a semi-permanent procedure. This statement is false.

Unless the pigment is implanted strictly within the epidermis (which turns over every 28 days), the implant is permanent. Unless their implants require a re-do every 28 days (as a result of complete pigment loss), the implant is considered permanent and, by extension, a tattoo.

The second layer of the epidermis is the *stratum lucidum*. These cells are clear and are unique to the palms of the hands and soles of the feet. This thick layer cannot be found on the thinner skin of the body, and can clearly be felt when comparing the skin on the soles of your feet to the skin on the top of the foot. The third layer of the epidermis is the *stratum granulosum*. This layer is known as the grainy layer and consists of flattened keratinocytes (cells that produce keratin). *Keratohyalin granules* are produced in this layer, which is a protein that, among other functions, fuses keratin filaments together. The fourth layer of the epidermis is the *stratum spinosum*. The cells in this layer are known as the "spiny cells" and consist of many keratinocyte cell layers.

Langerhans cells can also be found within this spiny layer.

The fifth and final layer of the epidermis is the *stratum basale*, also known as the *basal layer*, *stratum germinativum, the germination layer* or *the growth layer*. The melanocytes (melanin-producing cells) reside within this layer. Once basal cells die, they eventually float up to the surface layer (the stratum corneum). As subsequent cell layers die, they also find their way to the horny layer where retired cells are shed away in the 28-day cycle known as *cell turnover*. It is important to note that as the human body ages, the skin's cell-turnover cycle will become delayed and exceed 28 days. The 28-day cycle I mention is also based on healthy skin as opposed to someone with certain skin conditions (like *psoriasis*), that may have a hyper-accelerated turnover cycle. Just below the basal layer is the "plane of pure energy", where the epidermis meets the vascular layer of living skin known as the *dermis*. This layer is the vascular layer also known as the "live layer" of skin. This is the epidermal-dermal junction, the location where blood begins to flow, also known as the "sweet spot" in PMU. This junction layer is also known as the papillary layer of the skin.

The skin is composed of two compartmentalized teams that are fused together and work in perfect synergy. Like conjoined twins, the non-vascular *epidermis* and the vascular *dermis* form the skin. An example of the skin's harmony in motion is the physiology of touch. The superficial sensation of touch begins just below the epidermis. The skin is our largest sensory organ that, among other things, influences sexual attraction and sexual fulfillment. The skin is also an organ that is visually appealing, and by extension, visually stimulating. Like a fashion designer draping silk onto a dress form, skin is draped over the raw anatomy to create the most genius works of art created by nature. Skin is what defines the human form and accentuates its curves. It is what outlines the hills and valleys of the body's distinct contour. How many romance novels describe a woman's "silky soft skin" as an element of visual arousal that sparks the desire to touch her? As an avid reader of romance novels, I can assure you that the topic of skin comes up frequently! When the skin is touched in a tender way it feels pleasurable, especially when caressed consensually by someone you care about; physiological changes occur. When you touch the skin this way, you are actually stimulating hormones. Serotonin is the "happy hormone" that is released from the brain during touch. If you've ever experienced a relaxed state, you have experienced this phenomenon first hand!

This sensory feeling is a direct result of nerve endings, like *Meissner's corpuscles,* which are peppered throughout the dermis. Although *vellus hair* (peach-fuzz hair that is light in color) cannot be as easily seen as *terminal hair* (thicker, darker hair), it is responsible for some of the highly sensitive touch receptors that humans are programmed with. Hair is an extension of the skin, and it grows out of a vase-like structure known as the *hair follicle*. Although Meissner's touch receptors are more concentrated in the thicker portion of skin that is void of hair (the stratum lucidum), the hair follicle is directly responsible for the gratifying feeling of tingly touch that humans experience, as a result of the *arrector pili* muscle in action; this muscle lives just outside of the follicle, and is attached to the follicle wall. When the skin's nerve endings receive the sense of stimulation, it triggers the stimuli of the arrector pili, causing the hair on the body to act in a physical manner that we identify as "goosebumps". Since this muscle resides on the exterior of the follicle, it relies on the hair to manifest its sensory response to stimulation. In other words, hair adds value to the physical human experience, as well as the visual. In addition to hair, nerve-endings provide humans with a superficial feeling of pleasure; however, it is that goose-bump experience we get when pleasure is heightened that is directly related to the hair follicle.

The hair grows out of the follicle which is rooted within the cushy comfort of the dermis and extends up to the epidermis, before it exposes itself to the outer world (see Figure 3). Hair follicles are located within the skin, the point of which begin at the surface of the epidermis, leading down into live dermal tissue. Picture a glass test tube inserted within a container of packed, dense sand. Imagine that the surrounding sand is the skin (both epidermis and dermis) and the glass tube is a hair follicle. The center to bottom half of the container is packed with sand that represents the dermis, and the center to top half of the container represents the epidermis. Now imagine yourself inserting a sharpened pencil inside of the glass test tube, then gluing the sharpened tip to the base of the tube. This image represents a hair growing from the base of the follicle. The pencil represents a strand of hair, the test tube represents the hair follicle, the tip of the pencil represents the hair bulb (the living portion of a hair shaft), and the surrounding sand represents the skin. If you picture this image, you will note that the pencil (the hair) does not grow from the skin, but rather from the hair follicle itself, which, in turn, lives within the skin. The follicle resides within the skin but is not part of the skin's anatomy.

Clients often ask me if the tattooing process will affect their hair; I get this question all the time! They are concerned that the process will compromise their hair's growth pattern. When a client asks me this question, I first explain the process of tattooing: Pigment is implanted within the papillary layer of the skin (the dermis). The hair follicle is a separate entity and is not influenced by the action of tattooing. The hair follicle lives within the comfort and protection of the skin (dermis and epidermis), yet it is not part of the skin per se. Given that the skin experiences trauma during the tattoo process, it is not uncommon for hair to fall out during the scabbing and healing phase of the wound. However, tattooing causes temporary superficial damage to the skin and does not affect the papilla of the follicle, which manages the life of the hair shaft. In theory, the hair should grow back fine.

Consider the hair follicle to be a factory; each and every follicle is a hair-generating machine that operates independently from one another. As a licensed electrologist, I can tell you that it is not easy to terminate a hair follicle without damaging or scarring the skin unless the skin is severely compromised by major trauma. An example of severe trauma would be a third-degree burn or a severe laceration to the skin. When clients come to me for hair-removal, it can sometimes take years to terminate the life of the hair bulb from the follicle, particularly because electrolysis is only effective during the *anagen* phase of hair growth, the first of the three growth phases of hair. In electrolysis, the goal is to terminate the blood supply (the hair's source of nourishment) within the follicle while maintaining the integrity of the skin.

But I digress. There is a reason why I am discussing the anatomy of the hair follicle. Earlier I explained that the hair is an extension of the skin and that the follicle is one of the sources of sensory pleasure to humans. Why did I just explain Hair-Follicle 101 to you as it relates to pleasure? Figuratively speaking, I wanted to emphasize the layers of the skin. It serves many functions: It is a source of beauty, it is a source of pleasure, it is a source of function, and it is a source of pain. When a tattoo artist truly understands the eccentricity of the skin, he or she will become awakened to the mysteries that may have contributed to the undesirable artistic results of the past. An artist should not work on a living canvas they know nothing about.

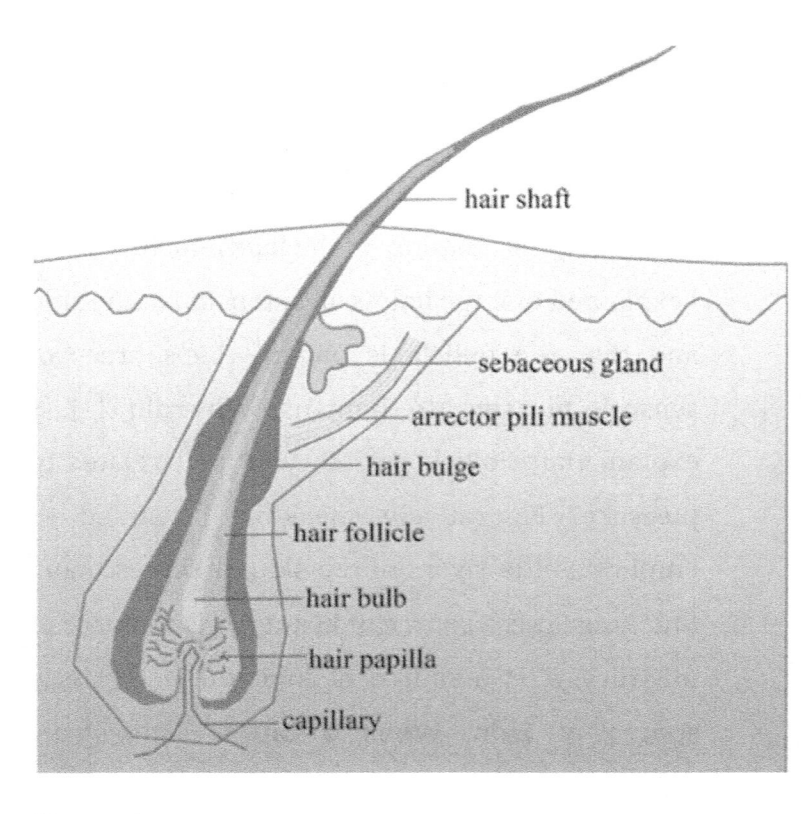

Figure 3

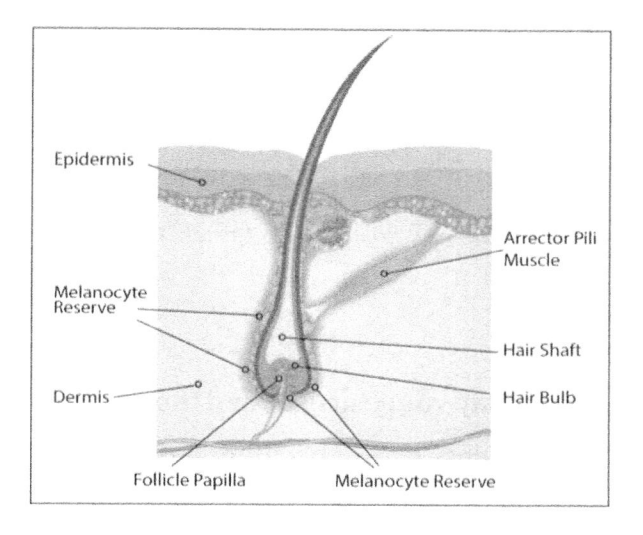

Figure 4

CHAPTER 5: DO YOU SPEAK FITZPATRICK?

The best color in the whole world is the one that looks good on you.

-Coco Chanel

I mentor many young artists, and they all possess the same desire to become exceptional at cosmetic tattooing. To achieve superior results, the tattooist must truly understand the eccentricities of skin. He or she must also learn how to speak the unique language of the skin, known as the *Fitzpatrick scale* and also known as "Fitz" or "Fitz type". Invented in the 1970s by Thomas Fitzpatrick, the Fitzpatrick scale was developed in an effort to determine how varying skin types respond to UV light produced by the sun. The Fitz scale is a numeric skin-type scale categorized into six levels. Please note that there are degrees in between each type; in other words, not all Fitz 3 types will look the same. One person can be a full-on Fitz 3, one can be a 3.4, and another can be a 3.9. There are levels within the levels, but here is the classic scale that all skincare professionals base skin-type on (see Figure 5).

The ability to identify Fitzpatrick skin types is a game-changer with regard to proactive tattooing. In my professional opinion, understanding the "what" of the Fitzpatrick scale will answer the "why" of unwanted melanin influence. If you look at the skin on your arm or your leg, you will note that the surface does not appear as white as a piece of copy paper; skin is alive and is not void of color. The skin is a living organ and has a color hue unique to you. This custom skin tone of yours is specifically based on your genetic makeup. For example, if you gave birth to a child, that child will possess 50% of your DNA. So if you had a child with a partner who is of Scandinavian descent with a much lighter skin tone and you are of Middle Eastern descent with darker, melanin-rich skin, that child may not exhibit any of your outward pigmentation at all and possess a

Scandinavian appearance containing more of a pheomelanin influence (blond hair and blue eyes). Although the child's skin appears to be lacking melanin, he or she is indeed genetically influenced by the melanin-rich parent, even if the child resembles the Scandinavian parent. In short, the child's skin is influenced by both melanin types and will respond to sun exposure differently than someone of pure Scandinavian descent with two blond, blue-eyed parents. This is why the PMU artist must ask probing questions pertaining to the client's ethnic origin. If the client is adopted, however, and is not familiar with their genetic background, it is up to the artist's trained eye to look for skin-clues (such as hyper- and hypopigmentation) to determine the most proactive protocol for the client.

If you are able to obtain information about the client's genetic makeup, it will be helpful when determining the hows and whats of the cosmetic tattoo procedure. "How will I execute the eyebrows for this client and what tools will I be using?" "What method of execution do I want to use on this client?" "What pigment shade will I use and why?" "What needle will I use?" What machine, what technique and what power level will I use, and why?" These are some of the essential questions you must ask yourself when analyzing a client.

As a permanent makeup artist, it is important to identify your client's ethnic ingredients in order for you to make smart choices pertaining to the tattoo procedure. Every individual will have a unique contribution of melanin (within their internal wound healing factory), and the amount of melanin, as well as the time that it will linger within the wound after it has healed, can never be predicted on first time clients. PMU is not one size fits all. Your decisions need to be mindful of the individual in front of you. Ethnic influence is essential information and will help you determine which pigment color (or combination of color-tones) to use on the client. What if you went by what your naked eye observed? You may tell yourself, "This person is blond with blue eyes and not very tan, so let me use the pigment I use for my fair-skinned blond blue-eyed clients" or, "this person has medium-toned skin, so let me classify her as a Fitz-3". These conclusions may set you up for failure, in terms of your healed work. What if the blond, blue-eyed client was a quarter Sicilian? The client's melanin influence (from her 25% Sicilian influence) may cause the blond pigment you chose, to heal darker and cooler than you had expected. This is why asking your client probing questions, about their ethnicity, is critical during the screening process.

The reason why melanin is relevant with regard to the micropigmentation process is that it will change the color of the healed result, and it can change the color of the healed permanent implant. This phenomenon is a result of the brown color tint that eumelanin sometimes leaves behind. In other words, you need to start thinking of melanin as an additional pigment color that may become implanted in the dermis, particularly for Fitz 4 through 6 types. For example, if you choose a color called "rose brown" and swatch that color on a piece of paper, it will look like "rose brown" on that piece of white paper. Skin, however, is a living organism and is not void of pigment, like white paper is. Unlike paper, the skin is loaded with color. Look at the skin on your arm right now. You may note color tones ranging from yellow, orange, red, brown, blue, and more. Human skin contains pigment; it is not void of color. In addition to the color that already exists within the skin, melanin will rush to the site of the injury (the PMU tattoo) during the healing process, which may compromise the implanted color-heal.

- TERYN DARLING'S DARK SKIN DISCOVERY

Educating oneself about the dichotomy of human skin, as it pertains to the PMU process, is a perpetual pursuit. "I only know that I know nothing" is the motto I live by everyday, which is why my unending quest for knowledge always leads me in the direction of industry leaders who have educated and inspired me along my PMU journey. I first reached out to Teryn Darling in 2015, before a live interview I had on Doctor Radio's *The Dermatology Show*, Sirius XM. I was to be interviewed by famed dermatologist Dr. Doris Day, as an expert within the micropigmentation arena. I panicked, anticipating that the questions from callers would be specifically about pigments and I was not well-versed, at the time, about the eccentricities of tattoo pigments. The difference between iron oxides vs. organic pigments was a topic that I needed to learn more about. I had been using the *Li Pigments* line for a while by then, and was impressed by the wealth of knowledge that Teryn offered to the industry, for free. I did not know her personally, but I was desperate to gain a few nuggets of pigment-wisdom before my live interview! So, I reached out to her in a desperate plea for a quick lesson. Let me just preface this by telling you that I was scared; Teryn was this powerhouse PMU boss who had, and still has, a strong presence about her. Why would this power-player take the time to speak to me, an unknown PMU artist who never even took a class with her? She is a pioneer within the PMU industry and I assumed that she would never give me the information I needed within the time I needed it; boy was I wrong. Not only did Teryn respond to my message that day, but she took the time to send me pigment charts, and data sheets, as well as a plethora of information that exceeded my expectations. As it turns out, the XM Radio interview (which was a success) never fielded questions pertaining to pigments, but that didn't matter. I was prepared, I was educated. I took the time to learn what I didn't know and Teryn, the teacher, was exceedingly willing to educate me without asking for anything in return. That always stuck with me. Her selfless desire to empower and educate me, propelled me to pay it forward. It is not only important, but it is imperative for industry leaders and educators to give back to hungry artists who truly want to perfect their craft. When I was thinking about adding a quote from a PMU educator about their perspective regarding melanin's influence on the cosmetic tattoo, there was only one person I reached out to. Teryn sent me 1,088 words in an effort to further empower the reader of this guide, however, I am the one more empowered than I was before I read her narrative. I am so thankful for Teryn, as well as all the industry educators who are just as generous and selfless as her. You are my heroes!

"I remember the first time I tattooed a set of eyebrows on a woman of color, and how that started me on a journey of discovery. She was a Fitz 5 and the first thing I experienced, as I was working on her brows, was that I could barely see the color, after my first pass or two; it just did not show up like it did on the lighter skin types. Finally by the third pass, the color had oxidized enough to where I could see it pretty darn good, but it appeared brown. I was using a soft black but it appeared brown in the skin and it seemed too light; just the opposite of the lighter skin types I had worked on in the past. When I saw her a few weeks later, I was looking at a stark, cool, ashy black tone as the healed result. She initially left with her brows looking lighter than expected, but they healed darker and ashier than I wanted. I was horrified. What happened? I spent years educating myself so I could answer that question and prevent that from happening in the future. In retrospect, it may have been the result of me overworking the skin. Although I was using a soft black pigment, I could barely see it the first two passes. By the time I was done, it still didn't look black, it looked brown. That really messed with me and may have caused me to overwork the area in order to see the depth of color that I thought I should be seeing. As artists, we are used to PMU looking TOO dark when client's with lighter skin types leave our office, so I thought it should look that way on the darker skin types as well. I was wrong. Overworking the skin was not the only contributor, it was also a result of the client's cool skin, hyper-pigmentation and the cool pigments I used.

Rose explains hyper-pigmentation so eloquently and in a way that is understandable, so I will just leave that with her. I would like to address the topic of pigments and color theory a bit: Dark is cool. The darker the skin, the cooler the skin. The darker the pigment, the cooler the pigment. Clients with darker skin and darker hair require (or request) our darker pigments, therefore, we are implanting cool into cool which can cause ashy, muddy or cool color results. If you combine cool into cool, and then factor in what hyper-pigmentation can do to the skin, your healed results may get you into trouble. When it comes to choosing pigments for the darker, cooler skin types, we must use darker, cooler pigments. There is no such thing as a "warm dark" pigment; it doesn't exist and is scientifically impossible to formulate. With that said, some pigment manufacturers, like *Li Pigments*, do add some warmth into the base of its cooler colors in order to help prevent ashing, and that can be enough with some darker, cooler skin types in achieving an acceptable healed-color result. However, there are many darker, cooler skin types where that will not be enough, and the technician will have to manually add some warmth to the mix. This is why *Li Pigments* has spent 25 years studying the different skin types, the healed results and have formulated a set of warm modifiers to suit all situations. It is important for the technician to develop their eye for the different skin types and undertones, to understand them and identify when to manually add some warmth to the pigment cup, and when not to. It is as equally important to add some warmth when needed, as it is to NOT add any additional warmth when it is NOT needed. In other words, as much as warm modifiers can be beneficial (and the reason we achieved an amazing healed-color result), when overused or misused, additional warmth can also be the reason that your procedures heal or age too warm.

In the twenty years since my first woman of color, I have studied and learned all the different undertones and overtones within the Fitz 4, 5 and 6 skin types. I have learned when I must add additional warmth and when it is not needed. When it is needed, I reach for either *UnGray* (orange) or *Gray Vanish* (Yellow-Orange) by *Li Pigments*. Which one I use will be dependent upon all the variables: How dark is the skin, what is the undertone revealing and what do I see with the overtone. There are different degrees of cool, even within the darker, cooler skin types. Learning the skin types, the undertones, the overtones and how to effectively counter their additions to our healed results (by using modifiers), really empowered me as an artist and nowadays I nail that healed-in color result every time. I also learned to not overwork the skin, it will just look lighter than expected when working on darker skin types.

By learning and understanding our canvas, it is possible to achieve dark, rich healed-in color, even on the darkest, coolest of skin types. I think that some artists are looking for a magical pigment line that requires no thought, no additions by the technician, and will heal beautifully each and every time. That pigment line does not exist. As tattooers of face we must invest our time in learning and understanding all aspects of the skin and all the variables of the skin that contribute to our healed results. Along with the study of the skin, it is important to understand how each needle configuration works, and how moving that needle in the skin contributes to our healed results as well. We must understand our pigments and how to use them properly. It can be daunting to fully understand and grasp all these outside influencers and how they affect our healed work, but once we do we can now predict the healed results with a great deal of accuracy. There is no better feeling than having that kind of control on your healed work.

I want to personally thank Rose for asking me to contribute a little something to her book. I was humbled when asked and excited to do so. This is a topic that we both care about passionately and people like Rose who take the time to put all their knowledge down on paper (for fellow artists) help grow this industry into what it is today, and will continue to become." -Teryn Darling, CPCP, is a world renowned PMU artist with over 20 years experience. She is the owner of Girlz Ink Permanent MakeupStudio in Las Vegas, Nevada, founder of Girlz Ink Training Academy, co-founder of America the Beautiful PMU Conference, Master Li Pigments distributor and co-creator of SkinFul Beauty Pigment Seal. Teryn is a licensed permanent makeup artist, a licensed tattoo artist, a licensed esthetician and an international trainer. To learn more about the study of color-theory and pigments as well as her elite educational academy, visit her at *GirlzInk.com.*

- MELANIN IS AN ADDITIONAL PIGMENT

If you view melanin as an additional pigment rather than simply a so-called "skin protectant", you will set yourself up for PMU success. You must understand that melanin will be added to that "rose brown" (I mentioned earlier) during the healing phase of the tattoo and may never leave the implant, thus altering the color. "But rose brown is beautiful, and it does not need melanin brown added to it!" you might say. Regrettably, it is the nature of the skin-beast, and melanin influence is a hand that all tattoo artists need to learn how to play. Is it possible that melanin will completely leave the site of the injury once the skin has fully healed? Yes, indeed, anything is possible! It's important to note that not all skin within the Fitz 4-6 categories will become affected by post inflammatory hyperpigmentation as it pertains to the trauma caused by PMU. However, you will not know this until the initial heal, and the probability that it will become influenced is very high, particularly in higher Fitz types..

The art of speaking Fitzpatrick is the most important language you can master as a tattooist, in my opinion. It is important to mention that most of us live in a melting-pot society; we live amongst individuals of mixed Fitzpatrick levels. If you encounter a client of Fitz 4 and Fitz 2 descent, for example, I recommend treating them as if they were a Fitz 3 or even 4. Always err on the darker side of their ethnicity when treating their skin.

SKIN TYPES	
Fitz Type 1	Very pale, white skin. Never tans in the sun, will always burn.
Fitz Type 2	White skin. Does not typically tan in the sun and will burn.
Fitz Type 3	Light brown skin. Has the ability to tan but will typically burn in the sun.
Fitz Type 4	Moderate brown skin. Sometimes known as "olive skin". This type tans easily in the sun and rarely burns.
Fitz Type 5	Dark brown skin. This type tans very easily and quickly, and will rarely burn in the sun.
Fitz Type 6	Deeply pigmented dark brown to blue-black skin. This type will never burn in the sun.

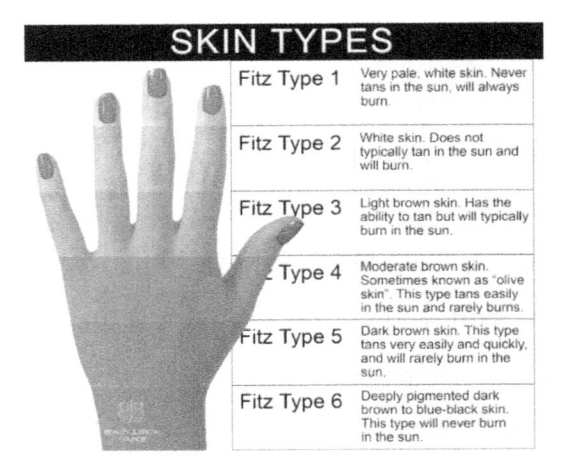

Figure 5

FITZ TYPE 1: This skin type is very pale white Caucasian skin and can include albino. This skin type is highly sensitive and can appear transparent on certain parts of the body. Fitz 1 hair is blond or red. This skin type can possess, but is not limited to, freckles. Fitz 1 eyes are blue or green and never brown. This skin type is extremely sun sensitive and will never tan; will always burn. Melanin: This skin type is only influenced by pheomelanin.

FITZ TYPE 2: This skin type has fair, white skin. Some ethnic examples of Fitz 2 types are Scandinavian, Irish, or Northern Europeans. The hair is blond, red, or very light golden brown, and the eyes are usually blue, green, gray, or hazel. Fitz 2 skin does not usually tan in the sun and can be expected to burn if not UV protected. Melanin: this skin type will have a combination of pheomelanin and eumelanin.

FITZ TYPE 3: People within this category have light brown skin. The ethnicities within this type can include some Middle Eastern, but typically European and Asian. Fitz 3 hair is chestnut, dark blond, or brown and the eyes are brown, blue, green, hazel, or gray. This skin type has the ability to tan but will burn in the sun if left unprotected. Melanin: this skin type will have a combination of pheomelanin and eumelanin.

FITZ TYPE 4: People within this category typically possess moderate brown skin. Fitz 4 types are sometimes known as "olive skinned" and can include Middle Eastern, Latina, Pacific Islanders, East Asia and some Indian. Fitz 4 hair is medium, or dark brown and the eyes are brown or hazel. This skin type can sometimes be an enigma. It is not uncommon to see a Fitz 4 Middle Easterner with hazel eyes and a lighter looking skin tone that may resemble a Fitz 3. In my opinion, Fitz 4 types are the most challenging to service with regard to PMU because of this. Fitz 4 skin will tan easily in the sun and will seldom burn.

Melanin: this skin type may possess some pheomelanin influence, but is primarily eumelanin.

FITZ TYPE 5: Individuals with this melanin-rich skin type possess dark brown skin. Ethnicities within this category include African, South east Asia, Central America, and North and South America as well as India and Pakistan. Fitz 5 hair is dark brown or brown-black, and the eyes are brown. This skin type tans quickly and easily and will rarely burn in the sun.

Melanin: this skin type may possess some pheomelanin influence but primarily eumelanin.

FITZ TYPE 6: This skin type is a deeply pigmented dark brown to blue-black skin tone. This type is also known as melanin-rich. Ethnicities within this category include, but are not limited to, Africans and Aborigines. Fitz 6 hair is black, the skin is brown-black or blue-black, and the eyes are dark brown. This type will never burn in the sun; however, it is not exempt from

UV-causing skin diseases like melanoma or other skin cancers. Melanin: this skin type is influenced only by eumelanin.

A common misconception is to classify someone's Fitzpatrick skin type simply by visually assessing their epidermis. In fact, each time I consult with young artists, the topic of Fitz type is usually an enigma to them. When identifying a client's skin type, I typically hear artists say, "This client is tan and has dark brown hair and dark brown eyes; therefore she is a Fitz 4." This is not the case for all skin in the Fitz 4 category; not all Fitz 4 individuals are tan. Fitzpatrick type is based on how the skin responds to injury, specifically the sun. Earlier I mentioned "varying degrees" within the Fitz scale. This is why two different Fitz 4 women can be classified within the same color-box, even though one may have visually lighter skin than the other; Fitzpatrick type has nothing to do with the exterior color of the skin. How does the client's skin respond when it is exposed to the sun for an extended period of time? Will she tan easily? Will she burn in an instant? Will she tan a little bit but also have a tendency to burn? Does she possess any skin discolorations, like brown spots on her face, chest, or arms? These are questions that must be discussed with the client. Many individuals of Middle Eastern descent, for example, are visually light skinned. A newer artist may classify an individual like this as Fitz 3 or even Fitz 2 based solely on their outward appearance! This is why probing questions about the client's origin must transpire. A fair-skinned Middle Eastern client will have a significant melanin response to injury, for the most part. A client with light skin, who is seemingly a Fitz 3 for example, may really be a Fitz 4. Again, it has everything to do with how she responds to injury and her genetic predisposition, her genetic makeup. This information is crucial because the PMU process does indeed cause an injury. Like pleasure-pain, the injury is meant to create something beautiful but is a physical injury nonetheless.

Let's say, for example, that a client walks into your office and is physically light skinned; in fact, if you had to guess, you would say that this person appears to be of German descent (blond, fair-skinned, light eyes). After you interview this client and ask questions about her heritage, you discover that her father is from Belgium, which is where she gets her golden hair, but her mother is from Egypt, with descendants in Morocco as well. This information is a game-changer! The client is visually Belgian, but since her genetic predisposition has classified her in a more melanin-rich category, you must treat her as if she is a Fitzpatrick 4 individual who has the genetic predisposition to hyper pigment despite the fact that her skin is light and she has golden hair. If your client has a genetic influence that will affect her skin's sun response, you must treat her as if she is of the higher Fitzpatrick category within her genetic soup.

Here's another example: If the client is three-quarters Irish and one-quarter Greek, you must err on the conservative side and expect the Greek influence to peek through during the healing phase. Can the client heal like someone who is of pure Irish descent? Yes, of course. However, if you assume that the client will pull toward a cooler tendency due to that Greek influence, she will seldom heal cool. If the client heals too warm, then it is an easy fix during the touch-up session. In other words, always assume that the mixed-race client will heal cool, even if they are outwardly fair skinned.

If you select a light brown pigment to swatch on your client's forehead, for example, you must be certain that the undertones of the pigment are warm, as in *golden yellow,* in an effort to negate the cool tendency within her genetic ingredients. In other words, the warm yellow orange undertones of golden yellow will cancel out the anticipated cool color pull when the skin heals. If you have to choose between a light brown and a dark brown, always err on the side of caution and conservatively choose the lighter brown tone if you are uncertain of the client's melanin influence. As I mentioned a moment ago, the artist can always make simple adjustments at the touch-up session to tweak the color, if needed. North America is a melting pot of races, and as such, genetic soup can come in many flavors. Ethnic influence should always be considered, and the dominant ingredient in the particular recipe should always be thought of as the dominant color-influencer.

Have you ever met with a new client and wondered how you can identify and/or predict her skin's response to injury? How can you predict if the client is prone to *post-inflammatory hyperpigmentation*? Look at her arms, her neck, and her face, are there any dark marks on the skin? What about dark circles under her eyes? If she does have dark under-eye pigmentation, how dark is it? What color is the under-eye area? Are the circles reddish-orange, dark brown, purple-black? Look closely at her skin, do you see any small white spots (hypo-pigmentation) on her face or arms? For the record, not all melanin-rich skin types (Fitz 4 through 6) will hyper pigment or possess dark circles under the eyes. Some do not have dark circles but will hyper pigment very easily. Moreover, some Fitz 3 types will hyper pigment, even if their skin tone is medium toned and not considered to be as melanin-rich as Fitz 4 or 5. In other words, when an artist consults with a potential client and conducts a skin analysis, he or she must look beyond the outward appearance of the client. When analyzing a new client, their skin should be an enigma to you until you unlock a few clues that will help determine the protocol best suited for them. Even if the client is fair-skinned, the response to an injury, like a tattoo, will be determined by the genetic cards that he or she has been dealt. As I've mentioned a few times already, the skin's physical response to wound-healing is influenced by genetic predisposition. Why have I repeated this concept so many times? The answer is, because it is the most important component when managing the tattoo, specifically on the eyebrows and lips. If the artist is uncertain or unclear about the client's response to the UV radiation generated by the sun, it would be prudent to select a pigment that is lighter and warmer than he or she would expect to select.

CHAPTER 6: WHY DOES POST-INFLAMMATORY HYPERPIGMENTATION MATTER IN TATTOOING?

I have so many liver spots, I ought to come with a side of onions.

-Phyllis Diller

Post-inflammatory hyperpigmentation (PIH) is the process associated with the skin's inflammatory response to injury, specifically subcutaneous injury. Trauma to the skin will ignite the inflammatory cascade, which stimulates melanocytes and melanogenesis. PIH is a phenomenon that can be observed when wounds darken and remain dark after they have healed. Have you ever noticed why some people, particularly those that spend a great deal of time in the sun, have brown spots on their face, arms, chest, and back? This post-wound stain is an enigma and can linger on the skin for weeks, months or even years. PIH can occur after an injury compromises the skin's vascular or nonvascular levels (dermis or epidermis). In other words, a healed pimple can hyper pigment just as easily as a deep gash to the skin. A cut with a kitchen knife, a burn on the arm, a bee sting, and a tattoo are all varying examples of injuries that can leave PIH in their post-healed wake.

The systems behind this response are not fully understood. It may, however, involve the activation of melanocytes as a result of inflammatory mediators, or *reactive oxidative species* (*ROS*) released by damaged skin. ROS are chemically reactive chemical-species containing oxygen. During times of environmental stress, such as UV radiation or heat exposure, ROS levels can increase dramatically. This may explain why those who suffer from *melasma* experience superficial brown skin-staining even after spending time in the shade. Melasma is a type of hormonally triggered hyperpigmentation that manifests as dark patches on the face. This skin staining response is triggered by heat as well as UV radiation. Exposure to heat will increase the ROS response even when the skin is in the shade!

The inflammatory response is also a result of the engulfing action of the *macrophages*. The word "macrophage" comes from a Greek root that literally means "big eaters." A macrophage is a white blood cell and has the ability to locate and gobble up foreign intruders that may harm the body. These cells are essential in wound healing, where bacteria and pathogens can invade a vulnerable open wound. Like little yellow Pac-Man gobblers, macrophage cells engulf foreign matter in an elegant wound healing dance. Macrophages are diverse cells and are one of the rock stars of the integumentary system. They are actually part of a team known as "antigen presenting cells", which are immune cells that referee the cellular immune response. The previously mentioned *Langerhans cells* are also members of this immune response team. After they complete their duty, the macrophage cells retain melanin in the dermis until the cells (and melanin) are broken down, enabling the melanin-pigment to linger within the dermis post-injury; post-inflammatory state. This is the brown healed landscape of skin that we call hyperpigmentation (HP). HP is a direct result of the injury's post-traumatic stress on the skin. PIH is when melanin lingers in the healed skin, post injury. HP is also an occurrence that results in a dark staining of the skin, but is caused by trauma that does not compromise the skin's barrier. Here are three examples of scenarios that cause HP without compromising the skin barrier:

1. Exposing the face to the sun after drinking a juice containing lemon can hyper pigment the upper lip area, unless a straw is used to drink from the glass.
2. Spending time outside without sunscreen, exposing the skin to UV radiation without broad spectrum protection, can cause HP

marks on the face and body that some may call "freckling".

3. Smoking asphyxiates the skin and introduces free radical damage to the skin by compromising the skin-cell's DNA. As a result of this free radical damage the color tone of the skin, particularly in the face and lips, can HP into a gray and/or grayish-brown tone.

When a wound is created during the process of tattooing the eyebrows, for example, melanin rushes to the site, in an effort to protect the traumatized eyebrows while they heal. In Chapter 3, I gave an example of a suntan, which is the most obvious melanin response pertaining to a wound, since the physical look of a suntan is actually a wound. When an individual is melanin rich and classified within the Fitzpatrick 4, 5 or 6 range by definition they have an elevated likelihood of experiencing PIH. This brown coffee or tea-like stain on the skin can be as small as a one-millimeter dot and can vary in size to a 15-millimeter splotch on the cheek. Regardless of the size, most people would consider a pigmented mark as an unwelcome resident on their body (especially on the face). If an individual is prone to hyperpigmentation, it would be impossible to predict the amount of staining their skin may receive after their skin is compromised. In other words, if a person is genetically predisposed to hyperpigmentation and is vulnerable to staining (as a result of any injury to the skin, regardless of how superficial), the chances of that individual becoming affected by HP is very high. After the skin experiences a wound, melanin can linger around the area even after the skin has healed and it is impossible to predict how long the melanin pigment will linger and if and when it will go away. It is extremely challenging to control these post-wound-stains. This is because there are two types of hyperpigmentation: *epidermal* (superficial) and *dermal* (which is deeper and much more difficult to manage).

Some individuals in the Fitzpatrick 4 to 6 range will not be affected by post-injury hyperpigmentation at all, and some individuals are hypersensitive to post-inflammatory staining. To reiterate, hyperpigmentation is based on how an individual responds to an injury; this response is influenced 100% by their genetic predisposition. Since genetics is a card that must be played in the game of skincare and tattooing, the artist must always work with the assumption that the client's skin will become impacted by excessive melanin. A colleague of mine recently reached out to me and asked if hyperpigmentation was a contraindication for tattooing. The answer is no. Once again, think of a suntan. That golden glow, although visually appealing, is indeed the skin's response to an injury. It's saying, "I don't want those harmful UV rays to burn me, so let me send my troops to the surface to protect myself." The troops I am referring to are melanin, which was sent to the surface by the melanocyte cells. In this example, a sunburn is the anticipated injury.

The more melanin-rich the individual is, the more melanin he or she has in their reservoir. In other words, a Fitz 6 has more melanin in their proverbial pigment glass than a Fitz 4 individual has in their glass. Compare a Fitz 1 with a Fitz 6. Fitz 1 has a limited supply of melanin, resulting in the appearance of a much lighter skin tone; the opposite is true for Fitz 6. A Fitz 6 individual has more melanin in their supply; therefore, their skin is much more vulnerable to PIH than a Fitz 1, who does not have the tendency to PIH at all. I personally view most of my clients as "skin that may experience PIH." If I treat them with this notion in mind, I will always work conservatively with regard to the pigments I select as well as the modality I choose to use.

CHAPTER 7: MELANIN INHIBITORS

Beauty is not in the face; beauty is a light in the heart.

-Kahlil Gibran

There has been a great deal of controversy within the permanent cosmetics industry with regard to tattooing on melanin-rich skin and lips. If you understand color theory as it pertains to makeup application, a pigment stain on the skin cannot simply be covered up with under-eye concealer. (The same notion holds true for under-eye discoloration.). You must first negate the color-tone of the stain using the appropriate complementary color-correcting shade (typically orange or salmon in tone). With traditional makeup application, the color correcting tone is first blended into the stained portion of skin, before the foundation is applied; the foundation should match the person's overall skin tone. The foundation should then be stippled over the color-correcting tone, like a flesh-tone blanket; this will create the illusion of an even skin tone. The operative word in the previous sentence is "illusion." Topical makeup is a game-changer for the woman with dark under-eye discoloration because she is now able to temporarily camouflage and brighten her unsightly under-eye discoloration.

Poorly executed micropigmentation, on the other hand, cannot simply be camouflaged with a proverbial pigment blanket; implanted color cannot simply be covered up like the under-eye scenario previously mentioned. Although some clients are indeed candidates for corrective color procedures, others are not. If pigment is heavily saturated within the skin, it may be in the client's best interest to physically have the pigment lifted with a laser or with the use of a hypertonic solution. As I've explained, the act of tattooing actually creates an injury and a wound cannot heal without the influence of melanin. Therefore, can melanin-rich lips, for example, be successfully tattooed? That decision must be made by an educated artist. The artist must determine if the client's melanin response to injury is low enough and light enough (in tone) to effectively treat the pigmented area.

I personally choose not to tattoo lips in my practice, so I decided to contact a colleague of mine, Lulu Siciliano, for her opinion on dark lip PMU. Lulu is the co-creator of *Evenflo Pigments* (developed by *Permablend Pigments*) and owns a permanent cosmetics practice in Miami, Florida, *PMU by Lulu. Evenflo* is a highly pigmented line of permanent makeup pigments designed specifically to neutralize dark lips. Lulu has dedicated her career to correcting melanin-rich lips through cosmetic tattooing and during my conversation with her, she showed me various photos of melanin-rich lips which she transformed into neutral tones. Lulu mentioned that she does not work on clients if the level of pigmentation within their lips is excessively dark; there is a limit to what she can correct successfully. She mentioned that her client's have realistic expectations and are willing to see any level of improvement compared to the dark lips they currently live with. There is no magic wand to neutralize melanin-rich lips and not every individual with dark lips is a candidate for this procedure, however options like a permanent lip correction can be game-changing for those individuals who suffer from the stigma of unsightly pigmentation in their lips. Lulu also added, "The art of the dark lip correction is not a new technique; I did not try to reinvent the wheel. In Russia, artists have been correcting dark lips for decades. In the United States, tattooing dark lips has been a debatable topic. However, the popularity of permanent makeup techniques, theory and technology have changed dramatically over the years. A dark lip corrective procedure can be life changing for the woman or man who lives with unsightly pigmentation in their lips. The only option that can help an individual like this is permanent makeup. I encourage all experienced PMU professionals to learn the art of dark lip corrections. It is an advanced procedure that requires extensive knowledge of color theory as well as a thorough understanding of the skin. I would not suggest attempting this procedure without thorough training".

If a client suffers (psychologically) from unsightly, pigmentation in their lips and contacts a reputable and experienced PMU artist in an attempt to create a better version of his or her lips, then I see no reason why the licensed and trained artist should not attempt to create some improvement. After all, the lips are the one feature of the face that typically cannot be covered up with makeup without having to reapply the makeup throughout the day, unlike other parts of the face where topical camouflage can last several hours. Moreover, a man who suffers from unsightly lip pigmentation is not likely to camouflage his lips with lipstick; permanent makeup via a "dark lip neutralization" is the logical option for him. If the client has realistic expectations and understands the risk that tattooing may create more pigmentation issues with regard to melanin's injury response, then who am I to question that? If, however, the lips are heavily pigmented to the level of nearly black (as with some Fitz 6 individuals), then it is not reasonable to assume that PMU pigment will cover up their pigmentation. I would, however, consider smoking a contraindication for a dark lip corrective procedure. As I mentioned a moment ago, smoking introduces the skin cells to free radical damage and will compromise the integrity of the healing lip tattoo. Once implanted, the pigment becomes part of the skin, but before it settles in, the color will become influenced and will change due to the inflammation cascade I mentioned earlier. The artist can not ignore the fact that melanin-rich skin is more at risk for the unwanted influence of dark pigmentation.

There is no way to negate the effects that melanin will have on healed PMU; however, there is a way to work with a client's pre-existing pigmentation if the tone is not too dark. Situations like this are not one size fits all and should be left for the experienced artist to decide.

Unfortunately, there is no magic wand to perform PMU miracles. In my opinion, a deeply saturated, melanin-rich lip that is dark brown to nearly black, is simply not a candidate for a permanent lip procedure. However, a pigmented lip may be a candidate for consideration. Not all people are candidates for all procedures, whether it is PMU, laser skin resurfacing, or even a chemical peel. Skin is not one size fits all and must be individually assessed, regardless of the modality being used. A woman who suffers from unsightly, melanin-rich lips has the right to take a risk and see where art (versus science) takes her. Art evolves when an educated artist, who has a thorough understanding of the skin and color theory, decides to push boundaries in an effort to make change. As an artist, an educator, and a student myself, I find inspiration when I see artists pushing the boundaries!

Regarding so-called "melanin inhibitors": For the record, there are many melanin inhibitors on the market, such as B-3 derivatives. One must consider, however, the client with hyperactive melanocytes (refer to chapter 3) since their melanin response cannot be predicted. Moreover, it is virtually impossible to predict who will possess overactive melanocytes. In theory, when a person uses a melanin inhibitor in the form of a systemic supplement, a topical serum or a cream, the prevention of melanin production can be slowed down or even halted. This can happen in two ways: The first way is to inhibit the production of melanin and yield the production of *tyrosinase*. Tyrosinase is an enzyme in the skin that, when blocked, will prevent the formation of melanin. It is an amino acid layer that is a component of melanogenesis and is needed for the production of eumelanin and pheomelanin. You can find tyrosinase inhibitors in over the counter skin care products containing hydroquinone, kojic acid, arbutin, and licorice extract, to name a few. The other way to inhibit the production of melanin is to obstruct melanin from actually entering into the melanocyte cells, to begin with. Studies show that niacinamide and soy block the transfer of melanin to the cells, which is why these ingredients can be found in over the counter skin lightening products. When a skincare product uses the word "lightening" or "brightening" in their buzz slogans, it means that it contains a melanin inhibitor. However, even if a melanin-rich client is using a melanin inhibitor, I urge you to be cautious and still anticipate melanin's potential influence on their tattoo. Just because the client is following a protocol that may protect his or her skin from pigmentation issues, this does not mean he or she will not experience an adverse reaction, making the problem worse. The takeaway here is this: Tread carefully on unpredictable skin. All skin types should be labeled "unpredictable" in my opinion; the takeaway pertains to each and every client that walks into your office. If you always err on the side of caution, you will seldom encounter errors.

CHAPTER 8: MELANIN'S INFLUENCE ON MICROBLADING

I will stop wearing black when they make a darker color.

-Wednesday Addams, The Addams Family

For the record, a gray heal can occur with a machine-tattooed eyebrow. Microblading (MB) does not render gray eyebrows. The action of microblading differs from a machine-implemented eyebrow, with regard to the modality used to create the tattoo. It is not uncommon to see a gray response after an eyebrow has healed regardless of the modality used. This has to do with the skills of the practitioner and the process utilized rather than from the actual tool used to achieve the pigment implant. The physical action of tattooing with a microblade causes a different type of injury to the skin as compared to machine implants; both modalities cause damage to the skin but in different ways. When the skin is compromised by trauma created by a laceration made by a microblade, it has to work more aggressively to heal, resulting in the possibility of scarring and hyperpigmentation. The reticular punctures created by a machine driven implant are not as severe as a laceration; however, this type of injury can also influence scarring if the artist does not practice proper stretching techniques as well as proper needle placement. Both modalities create a controlled injury, and both modalities require strategic skills to execute the art with minimal damage to the skin.

In the introduction of this guide, I wrote that *mastering the vast knowledge pertaining to the histology and the physiology of the skin, is essential to the PMU artist and just as important as practical PMU experience.* The popularity of microblading has manifested into a wave of new, enthusiastic artists learning the trade. As a result, some opportunistic trainers began offering vague curriculums with little regard to the integrity of micro-pigmentation is a complex art. New artists were left with lackluster training after paying a great deal of money to learn this new trade. This new wave was a revolutionary time for the permanent makeup industry! An influx of interest now appeared within a relatively small arena. Microblading, although an old modality, now put permanent makeup on the global map.

However, with that rise in popularity came some new-found problems within the process, including avaricious trainers attempting to compete with other trainers in an effort to dominate the MB landscape. The ego hinders creativity and growth; therefore, in order to become a superior artist you must put your ego aside and surrender to learning. As an educator, I was excited to see the growth within the permanent cosmetics industry, but I became concerned once I began speaking with new artists about the shoddy training that many confessed to having received. Supply and demand produced thousands of poorly-trained technicians across the world, many of whom were passionate artists in their own right, but who knew nothing about the art of MB. They invested their hard-earned time and money to learn a craft that would elevate their artistic career and assumed that the training they were receiving was top notch. Too many artists and not enough credible training resulted in mistakes along the way.

The good thing about mistakes, however, is that we can learn from them! If you are reading this guide right now, it means that you are investing in your own education in an effort to become better than you were. I personally live by the motto "I only know that I know nothing." This perspective is humbling and keeps me unlatched to change and growth. Although I have years of experience in the field of skincare and micropigmentation, I also learn something new every day, especially from new artists! This is a profound concept. Humans have the ability to make mistakes, learn from their errors and then master the task that was once unclear to them. The moment when information becomes knowledge is the moment when the artist becomes empowered and rewrites the narrative of their art. Isn't that what art is? Art is ever evolving and relies on the artist to move the needle that will empower other artists with the aim to pave the way for the next generation.

As I mentioned a moment ago, the physical process of implanting pigment with a microblade differs from that of the machine method. To reiterate, a machine-driven needle enters the dermis in a reticular fashion that creates micro-punctures, as compared to the microblading process which enters the dermis by creating an incision. Both methods achieve the same goal: They both create an opening within the epidermal/dermal junction of the skin, in a calculated effort to implant pigment. Once the pigment is deposited, it will reside within the dermal layer after the process of wound healing is complete. The difference between microblading versus machine implants is significant with regard to the physical process of entering the skin. It is that process that will ultimately compromise the integrity of the pigment as a direct result of the healing phase. This concept is what I detail in Chapter 6 of this guide. It is the skin's post-inflammatory response to wound healing that ultimately influences how the implanted color will heal. I will not go into detail regarding the process of wound healing, but it is important to understand that the manner in which a wound is physically created will have an influence on how the skin heals and how the skin will ultimately appear post-heal. If you'd like to learn more about the complexities of wound healing, I recommend reading "Principles of Infection Control for the Tattoo Industry" by Shanan Zickefoose, BSN, RN, CPCP (www.SPCP.org).

The act of entering the skin via an incision is much more traumatic than a reticular puncture created by a tattoo machine, if and only if the machine-driven implant is executed properly. The wound healing process is more aggressive when repairing a slice in the skin, as opposed to sealing a micropuncture. As a result of the trauma created by manually driven incisions, the skin becomes vulnerable to post-wound scarring, and the site of the wound will become open to hyperpigmentation impacted by melanin. As I mentioned, microblading is a complex procedure whereby the practitioner must use strategic stretching techniques in order to obtain a clean, precise channel within the skin. This process proactively prepares the wound for proper healing. Micropigmentation is a beautification procedure; however, as artists, we are indeed creating wounds within the skin to achieve that final artistic reveal.

Proper stretching, however, is not the only element for successful MB implants. Elite tools are critical in obtaining a precise micro entry within the epidermal dermal junction. Tools that are poorly manufactured cannot create a clean microchannel. Poorly made tools can create micro-snags and irregular incisions within the skin that will compromise the success of the implant. When a micro-channel is created in an elegant way, the skin will heal elegantly. However, when a micro-channel is created in a substandard manner, the skin will heal inadequately, resulting in the delivery of more melanocytes than expected (which translates to more melanin!). When viewed with the naked eye, substandard micro-channels can appear to look clean; however, when observed under a microscope, micro-strokes created with low-grade tools will affect the efficacy of the implants and will ultimately comprise the final skin-heal, as well as the color-heal. Thus the elusive "gray funky heal" as artists like to call it. As I have repeated many times throughout this guide, the practitioner must also have an elevated understanding of melanin's influence as it pertains to color theory, in addition to a general understanding of wound healing and how that relates to the histology and the physiology of the skin. If the artist practices within these microblading parameters, as well as analyzes and accounts for the client's genetic predisposition, it will set the final result up for great success, and the artist will be able to translate their artistic vision onto the skin's canvas.

CHAPTER 9: POST-WOUND PIGMENTATION AND THE MYSTERIOUS MELANOCYTE CELLS

The Merriam-Webster Dictionary definition of a wound reads: "an injury to the body (as from violence, accident, or surgery) that typically involves laceration or breaking of a membrane (such as the skin) and usually damage to underlying tissues."

Think about how many wounds you've experienced in your lifetime, as described by Webster. Although not violent, per se, a bug bite is a tiny wound that compromises the membrane (as Webster described). What about a tattoo? As artists, we do not perform surgery during the micropigmentation process; however, on a micro-level, the process is certainly a violent disruption of the body's armor and must be addressed by the skin cells during the healing phase of the wound.

After skin experiences post-wound healing, it can change into another version of itself. Sometimes a scar will form. Sometimes a change in texture can appear as well as a change in pigmentation. Once healed, the skin will function as it always had in the past with regard to its service as an organ; it will continue to serve and protect the body. However, on a micro-level, the integrity of the skin changes post-injury. Similar to a crack in drywall that is repaired with spackle and paint, skin will never go back to its original wholeness. So when considering future PMU touch-ups for the client, you must keep in mind the number of times the skin has been damaged. The skin is a living organism, and although it is self-healing, it's structure changes and becomes compromised over time if and only if it is continually injured, as in the action of repetitive tattoos to the same site.

Age is also a contributing factor with regard to wound healing. A person in their 20's has extremely viable stem cells within their skin reservoir, allowing the skin to continuously replenish the loss of collagen. After the approximate age of 30, this cycle slows down resulting in a decline in collagen production. The epidermis begins to become thinner as the skin progressively continues to age. This is why age is a valid factor when assessing the appropriate protocol for the client. The process of micropigmentation is not one size fits all. If the skin ultimately changes each time it recovers from an injury, the practitioner must make sure to reduce the number of PMU appointments on older clients. In theory, the skin of a 40-year-old is weaker than the skin of someone in their early 30's. Skin that is 50 years old is weaker than it was when it was 40 and so on. As the skin evolves over time, it loses its ability to bounce back with the tenacity it had in its 20's. The artist should create a PMU plan of action that is in the best interest of the client. If a client is in her 70's, for example, her touch-ups should be spaced out further apart as compared to a client in her 30s.

One of the most obvious changes in the skin, post-injury, is a scar. A scar can have a tremendous effect on the human psyche that, in some cases, can lead to social isolation. Close your eyes and picture the image of a scar. Perhaps you fell off your bicycle when you were a child, leaving you with a small pink scar on your knee, or maybe you underwent a facelift procedure that left you with tiny white scars behind your ears. When I suggest that a scar can lead to social isolation, you may, however, be thinking of another type of scar, such as a thick, keloid scar that is raised, red, and rope-like, that one typically sees after heart surgery or even a major car accident for example. These scars are difficult to hide and can influence the confidence of an individual who may be self-conscious about them. What about change in pigmentation, post-injury, without the skin's texture becoming compromised? Would you consider a change in the skin's color to be a scar even if the skin heals smoothly without any raised areas?

Merriam-Webster defines a scar as *"a mark remaining (as on the skin) after injured tissue has healed.... a mark left where something was previously attached..."* Okay, so Webster classifies a scar to be "a mark." If that is the case, would post-inflammatory hyper-pigmentation be considered a scar even if the skin heals smooth and without any raised areas? Let's think about that for a moment. If you looked at my naked face, it is peppered with unwanted brown spots and stains caused by melasma and post-inflammatory hyper-pigmentation. If it weren't for the glorious invention of makeup, I would, without a doubt, be locked up in social isolation rather than at my local coffee shop writing this guide for you at this moment! Simply put, when the skin is compromised in any way and has changed from its original glory, the result can be scarring, in many ways. Therefore, let me ask the question again: Would you consider a change in pigmentation to be a scar? Perhaps. If the skin becomes permanently influenced by any type of injury (controlled or uncontrolled) and that influence results in a change in either color, texture, or both, it is indeed considered to be a scar, which can, most definitely, have a scarring social impact on the client.

- THE MYSTERY OF THE FUNKY-HEAL

Let me throw a curveball at you. What if I told you that the reason why eyebrow pigment sometimes heals undesirably or "funky" (as some artists like to call it) is because of the actual hair follicle (see Figure 4), in addition to the skin that was injured during the tattooing process. The hair follicle may have as much of a melanin influence on the PMU heal as the skin itself. In Chapter 4, I went into detail describing the physical composition of the hair follicle. To reiterate, the hair follicle is a separate entity from the skin, despite the fact that it lives within the skin and among its surroundings. Although the follicle is not part of the skin's structure, it influences skin function. Think of the hair follicle as the skin's next-door neighbor who plays loud music. When the follicle-house is playing music, the skin-house feels the vibrations. Earlier I mentioned that the melanocytes are part of what I call the "melas team" of cells that reside within the epidermis. However, there is a significant melanocyte reserve that lives within each individual hair follicle, within the hair-bulb to be exact. This division of melanocytes operates independently from the melanocyte reserve that resides within the skin. Remember, melanin is the pigment that provides color to skin, eyes and hair. Although these follicular melanocyte reserves are assigned to the hair follicle, in an effort to infuse hair with color, how can we know for sure if this additional reserve will not influence a healing wound occurring on the skin, just a few millimeters from the hair follicle? We don't know, and PMU artists should be aware of its impending influence, specifically on Fitz 3 through 6 skin.

After reading this guide, we now know that the skin's wound healing response influences how the color of the epidermis will appear to the naked eye, post heal. We know why melanin rushes to the injury and how it affects healed PMU pigment. This phenomenon is caused by the action of the melanocyte cells which act as melanin influencers. When artists witness an unexpected change in pigment several weeks post-procedure, this metamorphosis is caused not only by the action of the basal-layer melanocyte cells, but also from the additional melanocyte cells that reside within the actual hair follicle itself. This is a game changing notion! This means that there is a melanin influence that may be propelled by two separate entities, as if one wasn't enough! This double-melanin response can affect the integrity of a PMU healed eyebrow and will explain why some artists may experience less than desirable post-heal results. Did you know that melanocyte stem cells are involved in the wound healing process? During wound healing, another breed of melanocyte stem cells (non-pigmented and void of melanin) migrate to the site. These cells are thought to be retired melanocyte troops, which are serving during the last leg of their service to the skin. This theory is somewhat of an enigma, but I wanted to mention them to pay homage to the mysterious melanocyte! After these non-pigmented cells sing their final song, basal-layer keratinocyte cells migrate to the wound during the third stage of wound healing, in an effort to assist with re-epithelialization, in an effort to bring *homeostasis* to the epidermis.

During wound healing, the process of skin regeneration to its original color tone is the fourth and final stage of the journey. If you have ever experienced a minor cut on the hand, you have observed the various stages of wound healing: the cut, the scab, the shed, and the repigmentation (Stage 1: Clotting, Stage 2: Inflammation, Stage 3:

Re-epithelialization and Stage 4: Remodeling). The repigmentation of a scar, as a direct result of hair-follicle-melanocytes, is dependent on a specific stage of hair growth at the time of the initial wound. As I mentioned earlier, there are three stages of hair growth: the anagen stage (growth phase), the telogen stage (resting phase) and the catagen stage (shedding/turnover phase). Hair follicle melanocytes actively undergo melanogenesis during the primary anagen phase of the cycle.

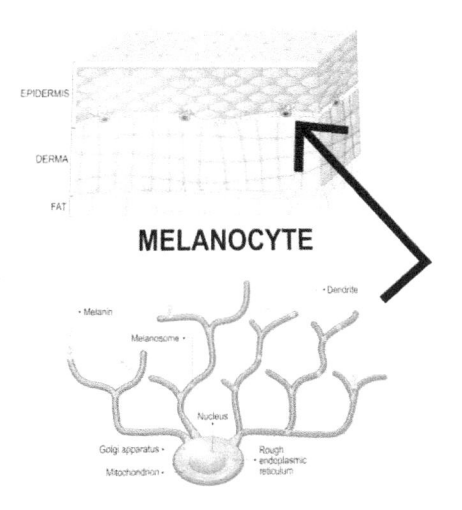

Figure 6

On the flip side, epidermal melanocytes have constant melanogenesis as a result of the perpetual influences, both intrinsic and extrinsic, that can affect the skin. In fact, the foundation of skincare as a multi-billion dollar stand alone industry is a direct result of those intrinsic and extrinsic influences on skin. An injury that occurs on terminal-hair bearing skin is more vulnerable to experiencing compromised re-pigmentation post-heal compared to a wound located on skin blanketed with vellus hair; an injury, such as a tattoo, that occurs on terminal-hair bearing skin, is more likely to experience post-inflammatory hyperpigmentation. This is a huge piece of information to digest. There are hair follicles on every inch of the human body, with the exception of the under-eye region, the lips, the palms of the hands and the soles of the feet. Although humans are covered in hair, most of that hair is *vellus hair*, which is the thin, lightly toned peach-fuzz hair that covers our bodies. Terminal hair is the thicker, darker, more robust hair that we see on our eyebrows, head, and lashes. Hence, if an injury (like a permanent makeup implant) occurs within hair-bearing skin, particularly skin covered in terminal hair, the skin is at greater risk of excessive pigmentation post-injury, as a direct result of the hair follicles.

I would conclude, however, that the more noteworthy changes in pigmentation occur within skin that is covered in dense terminal areas, like the eyebrows. Eyebrow hair follicles have a significant melanocyte stem cell reservoir within the bulb (see Figure 4). Wounds within this region are more likely to pigment than on the cheek, for example, where the follicles are vellus. It should not be lost on you that since there are no hair follicles on the lips, there is no chance of that "second melanin influence", although lips are indeed affected by melanin nevertheless. There are two opportunities where melanocytes can have a direct influence on a wound: that of the epidermis and that of the hair follicle. The takeaway here is this: The PMU artist must expect an eyebrow tattoo, for example, to become influenced by an unpredictable amount of melanin to the site as a direct result of the melanocyte's duty to serve the wound. The brows are covered in terminal hair follicles, and the skin's melanocyte activity may be heightened if, genetically, the client is predisposed to hyperactive melanocyte/melanin activity. Simply put, it's a crapshoot. There is no way of ever knowing how an individual will respond to injury and, moreover, how melanin will influence that injury.

An educated artist will know how to evaluate a client's skin type, determine if the client hyper-pigments (based on visual clues), and identify his or her ethnic origin (by asking the client probing questions). When choosing a pigment color for a client who is melanin rich and who shows signs of hyper-pigmentation, you must choose a color wisely and with caution; expect the color to change as a result of melanin's influence. A properly trained artist will understand that, in theory, melanin is an additional pigment added to their color-cocktail, no different than if they opened up a new dark-toned pigment, opened it up and added it to their pigment cup. This "new additional pigment color" (melanin) is not welcomed on our proverbial pigment tray, yet artists have no choice but to coexist with it. In other words, the melanin influence is out of your control, so always err on the conservative side when choosing a pigment color for your client, particularly at the first appointment session. Once the tattoo has healed, you can assess, note the color-heal, and tweak your protocols if needed. Warm colors layered-in lightly are much easier to manage in skin that is melanin rich, even if you think the execution is too conservative. It is much easier to add additional layers that are more robust, after the final healed result is assessed. Predictable is not a word that the PMU artist should use in their vernacular. If you work lightly, conservatively and use warm undertones, you will set the tattoo up for a beautiful end result.

CHAPTER 10: SKINCARE FOR THE HEALING TATTOO

Beautiful skin requires commitment, not a miracle.

-Erno Lazlo

The next book in my *PMU Useful Guide* series, *In the Skin*, details what the cosmetic tattooist should know about skin types, skin conditions, and how to care for the skin before, during, and, most importantly, after the permanent makeup procedure. Book 2 will provide you with a clear understanding of what the epidermis craves during the healing process in an effort to set the pigment implant up for continued success. I understand that many micropigmentation artists are not trained and licensed within the arena of skin health and skin care, so I am thrilled to change the narrative of the PMU industry by enlightening the reader about this topic! I am certain that *In the Skin* will spark some skin-related mysteries for the reader and answer important questions about the largest organ of the body as it relates to the PMU process.

As artists we are "image architects." As such, we should have a thorough understanding of the canvas for which we are manipulating. *In the Skin* will teach you how to assess the skin and identify common skin conditions, like dehydrated skin and oily skin. It will help you to create a post-PMU skin-care regime for your client that will set their tattoo up for long term success. An educated artist is an artist that the client appreciates, and adding value to your knowledge of the skin will make you a more credible professional. Please enjoy an excerpt from book 2 of my *Useful Guide* series, *In the Skin*:

There are a myriad of products on the market formulated specifically for skin growth and repair. Global sales for skincare products exceeded over $20 billion in 2019, more than half of that was sales within the U.S. alone! To say that the skincare arena is popular would be a tremendous *understatement. It is my professional opinion, as an aesthetician and as an electrologist who practices micro pigmentation, that if you set the skin-canvas up for wound healing success, the chances of PIH and scarring will minimize. Moreover, if you continue to nourish and protect the skin, post-procedure, by consistently using specific topical ingredients, the chances of color-retention increase dramatically. If you discern why the skin behaves the way it does, and if you understand what it needs to thrive, then why not institute a reliable protocol for your clients in the form of a proactive after-care regime? I like to think of PMU aftercare protocols as "skincare for the tattoo." PMU aftercare protocols should not, I believe, simply last a week or two, after the procedure. A well-executed post-tattoo skin care protocol should become the client's new normal every single day. Perpetual maintenance of the skin will render beautiful long term results for the tattoo.*

A tattoo changes the integrity of the epidermis. As a result, tattooed skin should be cared for every day in an effort to maintain the implant and keep the skin looking healthy. Skin requires balance and protection to survive and thrive. Skin is complex yet responds to simplicity. It maintains stability from proper hydration and acidic pH levels which, in turn, help to stabilize its protective barrier layer, the acid mantle. A healthy acid mantle can be a game-changer for lackluster skin. The acid mantle is the skin's barrier protectant, a super-thin, invisible cloak that is highly efficient in blanketing the stratum corneum and separating this outermost layer of skin from environmental elements. It is comprised of a unique cocktail containing sebum, dead skin cells, exocrine-gland secretions, and good bacteria in an effort to shield away viruses, pathogens, and bacteria that can otherwise harm the body.

The billion-dollar global skincare market I mentioned a moment ago creates products in an attempt to mimic the acid mantle! This barrier thrives in an acidic pH of 5.5 and one of its duties is to also prevent transepidermal water loss (dehydration). When the acid mantle is balanced, life on the skin planet is good. When, however, the acid mantle is disrupted, it is not balanced and the skin can show signs of dismay, which can manifest in the form of redness, peeling, itching, and even break outs. When the epidermis is not balanced, it has to work harder to obtain stability which provides wounds, like a tattoo, with an undesirable healing environment. One of the ways to maintain a balanced acid mantle is to sustain proper hydration and nourishment; the most effective way is to adopt a consistent skincare regime along with skin-friendly habits. Let me enlighten you on some proactive tips to help you care for clients with dry skin, in an effort to get their acid mantle in check, creating a desirable wound healing environment....

- DEHYDRATED SKIN

The skin is designed to regulate water balance. Dehydrated skin occurs when the epidermis experiences a loss of moisture; this can occur in oily skin as well. It is not uncommon to have oily-

dehydrated skin or dry dehydrated skin for that matter. Dehydrated skin has experienced transepidermal water loss (TEWL). This is when the skin is depleted of essential moisture as a result of evaporation. In dehydrated skin, the lack of water levels impacts the skin's function and regeneration, slowing down its ability to repair damage caused by trauma (like a healing PMU tattoo). Severely dehydrated skin can also have an adverse effect on the skin, leading to a loss of collagen and elasticity! In addition, the skin's natural exfoliation process is impaired, which results in a buildup of dead cell layers that clog pores and block the penetration of topical skin care products, like serums and creams. Imagine a window screen for a moment, those gray mesh liners that allow the outside air to pass through a window without debris or insects making their way into the space. In this example, the screen represents the skin and the thousands of tiny holes represent the skin's pores. When a window screen is crusted-over with dirt and accumulated debris, it is difficult for air to pass through. To improve the efficacy of the screen, you must scrub off the crust in an effort to expose the tiny air-holes, and allow the screen to breathe; the skin is no different. Skin that is left with an abundance of cell build-up is not a desirable environment for healthy PMU healing! If dry skin is left untreated resulting in an accumulation of expired cell-layers, the barrier function will become compromised and the pores will stockpile congestion; the end-result is semi-asphyxiated, dull skin. The goal is to create a successful post PMU healing environment, one that encourages proactive skin regeneration and encourages superior color retention....

- ASSESSING THE SKIN

Although the PMU artist is not required to become licensed in aesthetics, we are trained to ask the client about their skin type. As I mentioned a moment ago, there is a difference between a skin type and skin condition. Oily skin is a skin type, but eczema is a skin condition. Dry skin is a skin type, but psoriasis is a skin condition,...get it? An understanding of the numerous skin conditions (or at least the most common) would greatly benefit the PMU artist, in addition to understanding the different skin types. For example, have you ever met with a new client for a permanent eyebrow consultation and noticed severe flaking on the brow area? As you inspect their skin more thoroughly, you may note that the flaking skin is red and even oily in certain patches. The condition can also sometimes extend to the hairline and even around their nose. This can be a condition known as seborrheic dermatitis (SD), and I do see this a few times per year when I consult with clients. Seborrheic dermatitis is a superficial fungal condition that must be treated by a dermatologist. It is a form of dandruff and can affect the eyebrows and surrounding skin, causing redness, inflammation, and flaking of the skin. I would consider this chronic condition to be a contraindication for PMU, although SD can be managed with relative ease, many times by simply using a dandruff shampoo, recommended by their doctor, on the affected area. When a client of mine has seborrheic dermatitis and is managing the condition, I have a thorough dialog with them and alert them that the pigment implants can fade quickly over time and may become compromised by their condition. It is up to the practitioner to determine whether to treat the client with SD, as long as they provide the client with realistic expectations. Here are some tips that I use to help me determine...

- DRY SKIN VS. DEHYDRATED SKIN

As I mentioned a moment ago, it is not uncommon for skin to be both oily and dehydrated or even dry and dehydrated. To reiterate, dry skin references an individual's skin type and dehydrated skin is a skin condition. Dry skin is when the epidermis lacks a sufficient amount of sebum production and becomes deprived of the skin's natural moisture-producing oils. The lack of sebum production leads to TEWL because there is not enough of a lipid layer on the skin to prevent water from escaping. Sebum seals in moisture and without it, the skin can become dehydrated. Dehydrated skin can contain an abundance of congested hair follicles, due to excessive cell build up and can appear red and inflamed. Dry skin that lacks oil can feel itchy, appear flakey, and can even experience cracks that expose it to pathogens. Dehydration is caused by external factors in the environment that wick moisture out of the skin. For example, low humidity, as well as extreme wind in hot or cold environments, compromises the moisture levels within the epidermis. Dehydration can also be caused by foods that have a hypernatremic effect, such as diet soft drinks, caffeine, and sodium. An easy way to protect and hydrate dry skin is by applying a daily topical (emollient) humectant in an effort to moisturize the skin and seal in moisture....

Here's a dry-skin visual for you: Have you ever eaten bone-dry Thanksgiving turkey? You waited all year in anticipation of a juicy, savory bite until your culinary dream was shattered all because the chef didn't bother to baste it or bag it or whatever moisture-locking method cooks use to seal in the turkey's juices. This phenomenon also occurs when the skin is exposed to too much heat and/or excessively dry environments. Environments with little to no humidity will literally suck the moisture out of your skin. I live in South Florida, where one can almost drink the air due to the atmosphere's high humidity levels. This type of moisture is desirable for the skin because the epidermis requires water to thrive, and like a sponge, it absorbs water molecules from the atmosphere. But although humid, the South Florida environment is also hot. Heat will evaporate the water right out of the skin unless you occlude it with a moisture-locking barrier....

- LOCKING MOISTURE IN

A humectant attracts moisture to the surface of the skin while stabilizing the skin's barrier function. Like a sponge, it attracts water molecules in an effort to deliver hydration to the epidermis. There are a few types of humectants, and although thicker versions like glycerin are highly effective, many people do not care for the thick, heavy, sticky feeling that emollient-rich humectants leave on the face.....

You'll often find hyaluronic acid as an ingredient in over-the-counter products like eye creams, serums and face creams. Ingredients with low-molecular-weight hyaluronic acid are beneficial for deep hydration and a high-molecular-weight version will hydrate the skin's surface. By definition, hyaluronic acid has a larger molecule than an emollient humectant; the smaller the molecule, the deeper the product penetration. The protocol depends on the specific skin type in need of hydration....

Both types of humectants serve the purpose of hydrating the skin, it's simply a matter of what you prefer. Here's the rub, however: If you live in a dry climate, you must occlude the skin after applying a humectant. If you do not occlude the skin, in an effort to seal-in the moisture, the humectant you've just applied will work against the skin. Skin that resides in humid environments (like South Florida for example) does not necessarily need to be occluded unless it is excessively dry and/or if the skin is healing from a wound, like a tattoo. On the other hand, skin that resides in a dry climate with low humidity levels (like Arizona) cannot rely on the external environment to hydrate the skin.....

An occlusive will seal the wound, lock moisture in, and propel proper healing. Some examples of occlusive products are cocoa butter, beeswax, Aquafphor, and Vaseline. I am not addressing nutrition in this guide; however, it is important for me to note the benefits that nutrition has on the skin. Drinking plenty of water and eating water-rich foods can also help the skin maintain good water balance. If the skin is naturally lipid dry, it means that there isn't enough oil between the skin cells to lock moisture in. Dry skin does not have enough of its own essential oils between the skin cells, which results in transepidermal water loss (TEWL). In other words: without a proper lipid layer on the skin, water will escape and evaporate from the epidermis, so it must be sealed-in to prevent water loss....

The skin is more permeable when the barrier function is impaired; dry and/or dehydrated skin weakens the skin's barrier function. If the skin is dehydrated, it must be re-hydrated and nourished, prior to occluding. If dehydrated skin is not nourished and occluded, any product that you place on the face will simply absorb into the skin and disappear, like a tree branch thrown into quicksand. For example, have you ever applied makeup in the morning only to find that it has disappeared into the skin within a few hours of application? This may be a result of dehydrated skin that has absorbed the product into itself. Makeup is made to sit on top of the skin, so a proactive skin-prep that includes moisturizing and occluding will keep makeup resting on the surface of the skin. Dehydrated skin lacks water that impacts the skin's barrier function and regeneration, slowing down the skin's repair and cell turnover rate. This causes the skin's natural exfoliation process to become impaired, which results in a buildup of dead cell layers, like bricks piling up on top of one another. Restoring the skin's hydration levels will reinforce the skin's barrier function (the acid mantle), which is the holy grail of healthy skin. To hydrate the skin properly, I recommend....

FINAL FOOD FOR THOUGHT

And now, to ask the ultimate question: What is the secret to taking melanin out of the PMU equation? In Chapter 2, I stated, "The tattooist must be able to understand why melanin is produced and why it exists. This information is crucial for the successful longevity of a pigment implant." Isn't that the ultimate goal we, as artists, long to achieve? We spend thousands of dollars a year and countless hours in advanced training sessions for the sole purpose of the successful longevity of a pigment implant. Therefore, the answer to my question is that there is no answer. The melanin response of an individual is unique and unpredictable. In this guide, I gave you examples of various ethnicities and how they are influenced by melanin. The process of evaluating and assessing begins once a client makes an appointment for a cosmetic tattoo. After you review the client's paperwork to rule out contraindications, you determine his or her

Fitzpatrick type and begin to ask probing questions pertaining to their genetic predisposition. Once you are confident with the color and protocol you have decided on for their PMU procedure, you begin to execute your artistic interpretation as an artist. What if, however, you receive a call from the client after a month, informing you that the color is healing undesirably? Could it be that you misevaluated their skin? Could it be that the client forgot to mention a key ingredient in their genetic soup? Could it be that you failed to notice clues on their skin pertaining to hyper or hypopigmentation? Quite frankly, the skin is an enigma, and melanin's response as it relates to Fitzpatrick type is a mystery wrapped in a riddle. As long as you are educated and have learned as much as you can with regard to the skin's response to PMU-injury, then that is all you can do. At the end of the day, even the most seasoned artist will make mistakes, and mistakes are part of the learning process. An error can, however, be corrected or tweaked, and the ability and know-how to come up with solutions is, indeed, empowering! Happy tattooing!

GLOSSARY OF TERMS

Anagen: The first of three phases of hair growth. It is the growth phase of a hair shaft. This is the active stage of growing hair.

Arrector Pili Muscle: The tiny muscle that is attached to the exterior of the hair follicle-wall. It is a sensory muscle that provides "goosebumps" when the skin is stimulated.

Capillaries: A branching of tiny blood vessels. These tiny red or pink blood vessels can be seen on the cheek, nose, and eyelids.

Catagen: The second phase of a hair's life cycle. This stage is also known as the "resting phase". During this phase, the existing hair stops growing and starts the decline of its life cycle before it sheds from the follicle in the final "telogen" phase.

Couperose Skin: This is a condition where the skin, primarily the cheeks and the nose, are peppered with an abundance of tiny red blood vessels. I would classify this skin type as "sensitive", and it can easily become red when exposed to heat or cold or when the individual consumes spicy foods.

Dendritic Cells: Cells that directly support the immune system.

Dermis: The live layer of skin, located just below the epidermis. This layer is vascular and has a blood supply.

Dehydrated Skin: When the skin is lacking or depleted of water.

Dry Skin: When the skin is lacking or depleted of its natural essential oil (sebum), produced by the sebaceous glands.

Epidermis: The non-vascular portion of our skin. The exterior portion of the skin that coats our external body.

Epidermal/Dermal Junction: This layer is the curvy-border that distinguishes the epidermis from the dermis. This is the "sweet spot" layer of live skin where tattooists implant pigment.

Ethnicity/Ethnic Origin: The ancestral race of an individual.

Eumelanin: The dark pigment that most people possess to pigment their skin, hair, and eyes. There are two types of eumelanin: brown eumelanin and black eumelanin. When an individual has a small quantity of brown eumelanin within their melanocytes and an absence of black, this will give them the color known as blond as a result of the pheomelanin also present within their skin and hair.

Fatty Acids: Derived from plant or animal sources. Fatty acids are used in creams, lotions, makeup, and cleansers. They are emollient and can become comedogenic.

Fitzpatrick Scale: Also known as "Fitz type", "Fitz scale" or "Fitzpatrick type", the Fitzpatrick scale was developed in an effort to determine how varying skin types respond to UV light produced by the sun. This scale is used to identify a skin type based on a person's exposure to UV radiation. The Fitz scale is a numeric skin-type scale categorized in six levels, label one being the lightest skin tone and label 6 being the most melanin rich.

Hair Follicle: The tube-like structure that is found in the skin. This is what the hair grows from and is also the location of the sweat glands, the oil glands (sebaceous glands) and the arrector pili muscle.

Histology: The study of anatomy at the micro level. "Histology of the skin" is the microanatomy of the skin.

Homoeostasis: In a word, it means balance. When the skin is in a state of homeostasis, it means that all of its nuts, bolts and cogs are in perfect working order. The skin is a super-organ; self-healing and self-regulating. It's goal is to constantly be on call in an effort to create a state of perpetual homeostasis.

Humectant: A substance that brings water to the skin, after it has been applied topically;

humectant molecules attract water like a magnet is attracted to metal. Hyaluronic acid, glycerin, castor oil and honey are some examples of humectants. Not all humectants have an emollient effect. An example of this would be glycerin versus hyaluronic acid; glycerin is incredibly emollient, and hyaluronic acid is not. The job of a humectant is to keep the skin plump and hydrated by inviting environmental moisture to reside within the epidermis.

Hypo-pigmentation: A loss of color in the skin, caused by disease or trauma.

Hypernatremic/Hypernatremia: Dehydration (as referenced in this book) as it pertains to the skin, caused by the consumption of dehydrating foods or beverages, such as salt and caffeine. Humectants can also cause a hypernatremic effect if applied in an arid environment without an occlusive to seal moisture in.

Integumentary System: This system is one of 11 important organ systems of the human body. It encompasses the hair, skin, nails and exocrine glands. I mention this system because it directly applies to cosmetic tattoo artists.

Keloids: Keloids are overgrown scars that are raised and pigmented. This overgrowth of tissue happens when an abundance of collagen grows during wound healing, causing the skin to look bubbly, bumpy and raised. This is a genetic predisposition but is not harmful to the individual. You cannot simply remove a keloid because the act of removing it is an injury to the skin, and an injury is what stimulates the keloid to grow in the first place.

Keratinocytes: These are the primary cells that occupy the outermost layer of the epidermis. These are the receiving cells where the melanosomes deliver melanin.

Langerhans Cells: Immune cells that have dendrites. They reside within all layers of the epidermis, but primarily in the stratum spinosum. They are detective-cells that identify dangerous organisms that can harm the body. When the Langerhans encounter a harmful organism, they deposit them to other immune cells within the body where they are disposed of.

Lipid Layer: A lipid is a fatty acid, so a lipid layer is a fatty acid barrier that protects the skin.

Macrophage Cells: Also known as macrophages, they engulf foreign matter that may harm the body.

Meissner's Corpuscles: These are nerve endings in the skin that are responsible for sensitivity and the sensation of light touch, primarily on the palms of the hands and soles of the feet.

Melanin: The tint that gives skin its color; It is also part of the skin's defense mechanism and appears anytime the skin is faced with trauma, such as a cut, a burn, a but bite or when it is exposed to the UV radiation of the sun. Melanin is generated within the melanocyte cells.

Melanin-Rich: A melanin-rich person, also known as "skin of color," has an abundance of eumelanin within their melanocyte cells, with less or no pheomelanin. As a result, they possess a high percentage of the darker eumelanin pigment within their skin, labeling this type as "melanin-rich". This skin type is typically Fitzpatrick 4, 5, or 6. Melanin-rich can also be used to describe lips that are richly pigmented with a brown and/or black eumelanin tint. Fitz-1 types who are exclusively categorized as pheomelanin-influenced, would not be considered melanin-rich, this label is exclusive to the darker melanin skin-types.

Melanocyte Cells: These are melanin-producing cells that are located within the basal layer of the skin. Melanin is produced within this cell. The melanocytes directly support the skin's immune system.

Melanosomes: Pigment transport bubbles where melanin is produced. These bubbles are located within the melanocyte cells.

Melanoma: This is a cancer of the pigment-producing cells.

Melanogenesis: The creation /production of melanin.

White Blood Cells: These cells are immune cells, located in the blood, that protect the body from illness and disease.

Melasma: A superficial brown or coffee colored stain on the face. It's a form of hyperpigmentation, but is typically induced by hormones and can get worse when the skin is exposed to heat, blue light and UV radiation.

Occlusives: To occlude means to cover or seal. Common topical occlusives include caprylic/capric triglyceride, beeswax, petroleum jelly, and cocoa butter. These substances are hydrophobic, they repel water by creating a waterproof barrier. By applying a moisturizer with quality occlusives to the skin, the skin retains its water content. If applied regularly to the skin, in dry environments, occlusives on the skin reduce the rate of TEWL by over 90%.

Pheomelanin: Individuals with this type of melanin do not produce eumelanin. Rather than brown or black pigment in their skin and hair, the individual with pheomelanin has a red cast to their skin and hair. This skin will always be fair and sun-sensitive.

Physiology: The manner in which a living organism functions. "Physiology of the skin" refers to the normal function of the epidermis and dermis.

PMU (Permanent Makeup): Is also referred to as cosmetic tattooing, microblading, and micropigmentation. This is a semi-invasive procedure whereby a licensed Tattoo artist implants pigment, within the dermal layer of the skin, in an effort to enhance, camouflage or improve a specific part of the body where the client desires an improved appearance.

Post-Inflammatory Hyperpigmentation (PIH): PIH is the process associated with the skin's inflammatory response to injury, specifically subcutaneous injury. Trauma to the skin will ignite the inflammatory cascade, which stimulates melanocytes and melanogenesis. The phenomenon known as "post-inflammatory hyperpigmentation" (PIH) can be observed when wounds darken and remain dark after they heal.

Reactive Oxidative Species (ROS): ROS are chemically reactive chemical-species containing oxygen that is released when the skin becomes damaged. During times of environmental stress, such as UV radiation or heat exposure, ROS levels can increase dramatically.

Scars: A scar is when a mark remains on the skin after it has healed from a wound. This mark can contain an overabundance of pigment or it can be void of pigment. It can feel flat to the touch or it can be slightly raised as a result of an overgrowth of fibrous connective tissue.

Sebaceous Glands: Sebum-producing glands that are connected to the hair follicle, and use the hair-shaft for distribution of the sebum-oil it produces. This oil is what gives skin its moisture and is the main ingredient to support the skin barrier function.

Sebum: The oil that is produced by the sebaceous glands. This oil is the skin's built-in protectant that provides it with moisture.

Skin Type: Your genetic predisposition determines your skin type. Your skin is what you are predisposed to from birth. Although skin conditions (like acne) can also have a genetic root, skin types are more basic and void of a medical diagnosis, such as "dry skin" or "oily skin".

Skin Condition: Skin conditions can be genetic, such as acne, but can also become influenced by extrinsic and intrinsic factors such as climate and lifestyle choices pertaining to diet and hygiene. Dehydrated skin, for example, is a skin condition that can be fixed simply by drinking more water as well as topically hydrating the skin with humectants. Eczema, on the other hand, is a chronic skin condition that requires medical attention from a physician.

Stratum Corneum: Also known as the "horny layer". It is the outermost of the epidermis; the top layer exposed to the outer world. It is primarily comprised of non-living keratinocyte cells. This layer sheds every 28 days (on normal skin).

Stratum Granulosum: The third layer of the epidermis, known as the "grainy layer". This layer is comprised of flattened keratinocyte cells.

Stratum Lucidum: The second layer of the epidermis. These skin cells are thick and clear, and unique to the palms of the hands and the soles of the feet.

Stratum Spinosum: The fourth layer of the epidermis and one of importance to the PMU artist. The cells in this layer are known as the "spiny cells" and contain keratinocytes.

Langerhans cells can also be found within this layer. This is the layer where the melanocytes and melanosomes deliver melanin to the keratinocyte cells.

Stratum Basale: Also known as the "basal layer." This is the fifth layer of the epidermis and is connected to the epidermal/dermal junction. The melanocytes reside within this layer.

Terminal Hair: The thicker, darker hair that humans possess on their bodies, such as the hair on the head, eyebrows, eyelashes, and mustache/beard (on men).

Telogen: The third stage of a hair's life cycle. This stage is known as the "shedding/turn over" phase.

Vellus Hair: Also known as "peach fuzz" hair, this thin hair covers the human body, with the exception of the soles of the feet and palms of the hands. This hair is sometimes difficult to see because of its thinness and light hue. The hair that covers the forehead, nose, and cheeks (on a woman) are examples of vellus hair.

Wound Healing: Wound healing is an elegant repair process where the skin and the underlying connective tissue repair themselves after an injury.

ABOUT THE AUTHOR

Rose Prieto began her career as an eyebrow and makeup artist, nearly 30 years ago during her years as a Theatre Arts major in college. It was her fondness for character development that inspired her interest in stage make-up, which led to her passion for manipulating the contours of the face, using shadow and light. Rose is a licensed cosmetic tattooist, esthetician and electrologist living in Miami, Florida where she owns and operates a private aesthetic practice, *Beauty and Brow Lounge*. Her passion within the PMU arena ignited when she began mentoring young artists. It was her love of educating and empowering others that propelled her to write her first PMU educational guide, *How Fitzpatrick and Melanin Influence the Cosmetic Tattoo,* which she has turned into an online course of the same name. As a published beauty writer and content creator (since 2013), Rose has dedicated her career to educating others through her knowledge of skin care and beauty. She is currently developing the second book in her PMU *Useful Guide* Series, *In the Skin*, and continues to host her own weekly podcast, *The Beauty Lounge*, where she empowers listeners to become the best aesthetic version of themselves (tune-in wherever you listen to your favorite podcasts!). To keep the conversation going, please follow Rose on social media @Beauty and Brow Lounge and @The Beauty Lounge Podcast or visit her website at BeautyandBrowLounge.com.

Made in the USA
Coppell, TX
19 October 2021